T0140276

Human–Computer Interaction Series

Editors-in-chief

Desney Tan
Microsoft Research, Redmond, WA, USA

Jean Vanderdonckt
Louvain School of Management, Université catholique de Louvain,
Louvain-La-Neuve, Belgium

The Human–Computer Interaction Series, launched in 2004, publishes books that advance the science and technology of developing systems which are effective and satisfying for people in a wide variety of contexts. Titles focus on theoretical perspectives (such as formal approaches drawn from a variety of behavioural sciences), practical approaches (such as techniques for effectively integrating user needs in system development), and social issues (such as the determinants of utility, usability and acceptability).

HCI is a multidisciplinary field and focuses on the human aspects in the development of computer technology. As technology becomes increasingly more pervasive the need to take a human-centred approach in the design and development of computer-based systems becomes ever more important.

Titles published within the Human–Computer Interaction Series are included in Thomson Reuters' Book Citation Index, The DBLP Computer Science Bibliography and The HCI Bibliography.

More information about this series at http://www.springer.com/series/6033

Anton Kos · Anton Umek

Biomechanical Biofeedback Systems and Applications

 Springer

Anton Kos
Faculty of Electrical Engineering
University of Ljubljana
Ljubljana, Slovenia

Anton Umek
Faculty of Electrical Engineering
University of Ljubljana
Ljubljana, Slovenia

ISSN 1571-5035   ISSN 2524-4477   (electronic)
Human–Computer Interaction Series
ISBN 978-3-030-08232-1   ISBN 978-3-319-91349-0   (eBook)
https://doi.org/10.1007/978-3-319-91349-0

This Springer imprint is published by the registered company Springer Nature Switzerland AG
The registered company address is: Gewerbestrasse 11, 6330 Cham, Switzerland

Preface

A few years ago we started developing an application that would help Golfers Master a *perfect* golf swing by using an augmented biofeedback. We enthusiastically began our research, but soon found out that the literature in the area of biofeedback systems and applications is scattered across different research areas and domains. It was difficult to find books or articles relevant to the topics of our interest. A good deal of works was dealing only with a very narrow research topic or they were written as reviews and surveys. We did not find any works that would systematically and concisely discuss the research area of augmented biofeedback systems and applications; the area of our interest.

With the continuing research in this scientific area, we became aware that many of our research colleagues came across the same difficulties. An idea to write a book that systematically covers biomechanical branch of biofeedback systems was born. We made every effort to organize the content of this book systematically, clearly, concisely and in a way that reflects the most relevant topics in biomechanical biofeedback systems and applications from the view of engineers with expertise in signal processing, communication, and information technologies. The emphasis is on systems with augmented feedback and special attention is given to systems and applications that use technical equipment to provide concurrent feedback to the user.

The book starts with a relatively comprehensive introduction that tries to present the broad interdisciplinarity of this research area. Then its focus narrows to the explanation of biomechanical biofeedback, its systems, and its different implementation architectures. The second half of the book is dedicated to the applications of biofeedback systems in sport and rehabilitation. In connection to that we discuss some of the most notable problems regarding the performance limitations of the available technologies that are used in such systems. The last chapter is devoted to the detailed presentation of biomechanical biofeedback systems and applications developed and implemented by the authors of this book.

This book is intended for researchers who are interested in a systematic presentation and discussion of the most relevant and interesting topics of biomechanical biofeedback systems and applications. We sincerely hope that we succeeded in connecting the scattered pieces of knowledge about biofeedback systems and applications in a concise, well-organized, understandable, and readable text that fills the void in this research area.

Ljubljana, Slovenia   Anton Kos
July 2018   Anton Umek

Acknowledgements

We are very lucky, having such great wives and families, who have been patiently supporting and encouraging us during the long hours of writing of this book.

The book would have not been possible without the initial inspiration and continual support from Prof. Sašo Tomažič, who is actually *responsible* that our research interest turned toward biofeedback systems and applications.

The extensive research activities, which led to a considerable portion of this book's content, were in part supported by the Slovenian Research Agency within the research program *Algorithms and Optimization Methods in Telecommunications* (research core funding No. P2-0246).

Thank you all!

Contents

Chapter 1
Introduction

Biofeedback is a highly interdisciplinary area of research that combines knowledge of several scientific fields; from medical and social sciences to engineering and natural sciences. Biofeedback systems studied in this book include several elements; one of them is a person, a user of the system that presents its biological component. Other elements of the biofeedback system can be technical devices, persons, or a combination of both.

In a *biofeedback* system, person's (*bio*) body functions, parameters, and states are sensed (measured), processed, and relevant results are sent back to the person (*feedback*) through one of the human senses. The person tries to act on received information to change body functions, parameters, and states in the desired way (Kos et al. 2018; Sigrist et al. 2013).

The focus of this book is on technologies and technical devices that are used to provide sensing, processing, and feedback functionalities of biofeedback systems. For example, biofeedback systems with technologically augmented feedback can obtain information that is out of reach of human senses or the information that is beyond human senses capabilities. Such systems employ the expertise and knowledge from a number of research areas and fields of science. Some of the research areas, which are contributing to the interdisciplinary research of technologically augmented biofeedback systems, are:

- *Sensing*—chemistry, physics, biology, medicine, microelectronics, electrical engineering, mechanical engineering, nanotechnology, and others.
- *Processing*—mathematics, physics, computing, informatics, communications, electronic engineering, and others.
- *Feedback*—psychology, sports, physics, electrical engineering, communications, mechanical engineering, and others.

Biofeedback is a research area with a broad range of applications. This book is dedicated primarily to the subarea of biomechanical biofeedback systems as defined in Giggins et al. (2013) and discussed in Chap. 2. Biomechanical biofeedback systems are particularly useful in sport and rehabilitation, where they are used primarily for

© Springer Nature Switzerland AG 2018
A. Kos and A. Umek, *Biomechanical Biofeedback Systems and Applications*,
Human–Computer Interaction Series, https://doi.org/10.1007/978-3-319-91349-0_1

accelerated motor learning. Motor learning is itself a large interdisciplinary research area that encompasses a number of scientific fields, such as psychology, medicine, physical rehabilitation, sport, and others.

Throughout the book, special attention is given to the real-time biofeedback systems that provide concurrent augmented feedback. Such systems are the most challenging variants of biofeedback systems that are also the most complex and the demanding in terms of technology. Consequently, at the time of writing this book, implementations of real-time biofeedback systems are scarce, but they have great potential in various applications in sport, rehabilitation, recreation, and some other areas of human activity.

Biofeedback systems are usable in many areas of our lives. In this chapter we shed light on some of the aspects and backgrounds of human activity that demonstrate the wide applicability of biofeedback systems. We introduce the most important technologies contributing to its successful implementation and operation. We also present our vision of future development and use of various implementations of biofeedback systems.

1.1 Benefits to the Society and Individuals

As already mentioned in the first paragraph of this chapter, one of the elements of biofeedback systems is a person—the user of the system. Therefore the benefits of biofeedback for an individual should not be hard to assess, but what about the benefits for the society? In the following sections we present some of the most obvious benefits of biofeedback in different areas of human activity and from different societal aspects.

1.1.1 Quality of Life

It is natural that humanity strives towards higher quality of life (QoL) for each individual and for the society as a whole. In recent years, quality of life has become one of the main research topics in many fields of science. For example, when measuring the success of a country or a society, the QoL index is increasingly used instead of the gross domestic product. The QoL of an individual can be assessed through many aspects, such as material and physical well-being, health, family life, community life, political stability and security, job security, gender equality, and others (Tomažič 2006). Biofeedback can help raising the QoL particularly in the areas of health and physical well-being.

Health and physical well-being are without doubt two of the most important aspects of a person's QoL. Healthcare is also one of the most important service systems of a modern society. Healthcare aims to maintain or improve the health of the population and of the individual by prevention or treatment of injury, disease, and illness. In recent years, prevention has been given increasing preference over treatment. Leading a healthy lifestyle is the most straightforward way to prevent disease and illness. When discussing a healthy lifestyle, two factors are repeatedly mentioned: a healthy diet and an appropriate amount and quality of physical activity. While a healthy diet is not within the scope of this book, physical exercise, as a form of physical activity, is one of its main themes. Engineering has always played a major role in increasing the QoL (Tomažič 2006); therefore, we also strive to contribute our knowledge and expertise for increasing well-being and health. While biofeedback is also used in treating illness and disease, this book concentrates on biomechanical biofeedback. It presents a number of ways of how the use of biomechanical biofeedback system can help with improving the quality of physical exercise and injury prevention.

1.1.2 Health—Rehabilitation and Injury Prevention

Common knowledge, confirmed by a number of studies, is that physical activity contributes to good health. The results of *The European Youth Heart Study* have shown that children with lower physical activity levels have a higher cardiovascular disease risk factor (Andersen et al. 2006) and that high cardiovascular fitness is associated with a low metabolic risk score in children (Ruiz et al. 2007). There is also evidence that regular walking exercise has various health benefits (Hanson and Jones 2015) and that aquatic high-intensity interval training can be beneficially applied to clinical and healthy populations (Nagle et al. 2017). Additionally, a recent study from Lohne-Seiler et al. (2014) has shown that there is a strong correlation between the registered physical activity and *very good health* status in the elderly population. Some of the above studies would have not been possible without the use of (wearable) technology that is discussed later in this book.

Motivation

The primary motivation of our research activities is to develop and implement biomechanical biofeedback systems based on wearable device that would be used in applications for the assistance in injury prevention and rehabilitation in various fields of healthcare that are related to physical activity. This research topic includes nearly all of the previously described research fields.

Special attention should be given to applications that monitor and/or direct physical exercise, with the final goal of rehabilitation or injury prevention. Such applications can be successfully used in healthcare institutions that carry out physical rehabilitation programs (Gruwsved et al. 1996) or by individuals and organizations that practice recreational exercise or sport training (Silva 2014) where their role would be primarily in injury prevention.

Example—Swimming Rehabilitation

We have designed a biofeedback system with a waterproof wearable sensor device (Kos and Umek 2018). The system is presented in detail in Chap. 7 of this book. We have performed a case study of a rehabilitation swimming exercise. This case study investigates the possible benefits of using wearable sensor devices for the support of physical rehabilitation through swimming exercise.

Swimming is chosen because, historically, hydrotherapy is viewed as a central treatment methodology in the field of physical medicine (Becker 2009), and because numerous research papers show that aquatic therapy and swimming are the exercises of choice in numerous rehabilitation scenarios after injury (Nagle et al. 2017; Prins and Cutner 1999), surgery (Singh et al. 2015) and are beneficial even for people with autism (Yilmaz et al. 2004). In aquatic therapy and swimming, the water buoyancy reduces the effects of gravity on the body. This is particularly beneficial for the spine and joints such as the hips, knees and ankles (Prins and Cutner 1999). Swimming is recognized as an important physical exercise for rehabilitation, because various spinal diseases and injuries show through movement asymmetry during swimming (Becker 2009).

Review papers from Camomilla et al. (2016) and Neiva et al. (2017) present the results of numerous studies about the use of wearable sensors and technology in sport and physical exercise. In Camomilla et al. (2016), the results show that in cyclic sports wearable sensors are most frequently used in distance running and swimming. Neiva et al. (2017) have identified 603 studies published between 2007 and 2016 that discuss wearable technology for measuring physiological and biomechanical parameters. Out of 112 studies that focus on physical activity, 13 were about swimming. Authors of both papers conclude that wearable technology can be effectively used for monitoring physical activities, including swimming.

The presented concept of a biofeedback application for swimming rehabilitation in healthcare proves the suitability and usability of biofeedback and wearable sensors in swimming rehabilitation therapy.

1.1.3 Physical Well-Being—Sport and Recreation

Physical activity and physical exercise are becoming increasingly important aspects of our lives. It is a necessary and required ingredient of a healthy life style and there is no doubt that it contributes to our physical well-being and consequently to better quality of life.

While sport used to be a synonym for physical activity performed in a person's free time that might not be true anymore. We can roughly categorize the physical activity into recreational sport or recreation, amateur sport, and professional sport. Each of the three categories has a separate place in the society and includes people with different goals.

One thing is common to physical activities performed by people; the need and the urge for the quantification of their physical activity (McGrath and Scanaill 2013). It seems that the comparison, quantification, and competition are in the human nature. Sport and recreation are perfect mediums for all of the abovementioned. A direct comparison is the most straightforward and it needs no technical aid. For example, in the direct comparison it is easy to establish who runs faster, who throws further, who is stronger, etc. For the rest, we need some sort of quantification that is in great majority of cases provided by means of technology. For example, timing of a run, measuring of a jump length, monitoring of a tennis serve speed, etc. These technologies have been available for many years now and they are routinely used in training and competitions. For example, a simple manually operated stop-watch is routinely used for monitoring activities and progress during training in many sports.

1.2 Sport and Rehabilitation

At the time of writing this book, sport and rehabilitation are the most promising and the fastest developing areas of biofeedback use. This claim is supported by a large number of research projects and articles about biofeedback systems and applications in sport and rehabilitation identified in survey and review papers by Baca et al. (2009), Giggins et al. (2013), Huang et al. (2006), Lauber and Keller (2014), Neiva et al. (2017), and Sigrist et al. (2013).

1.2.1 Advantages of Biofeedback

The most notable advantage of biofeedback use in sport and rehabilitation is the possibility of accelerated motor learning. In rehabilitation the motor learning is focused on reinstating the normal or natural movement patterns after injury (Giggins et al. 2013). In sport the main driving force is in gaining any kind of competitive advantage over other contestants.

Science, engineering, and cutting edge technologies are being increasingly valued in modern sports. They provide new knowledge, expertise, and tools for achieving a competitive advantage. One such example is the use of biomechanical biofeedback systems that offer the possibility for achieving a competitive advantage over the traditional methods of motor learning and training. An important area of research is the use of real-time biofeedback systems supported by various motion tracking systems. The majority of motions tracking systems are based upon motion tracking sensors and sensory systems, such as accelerometers, gyroscopes, and various optical systems that are discussed later in this book.

1.2.2 Biofeedback Requirements and Success Conditions

The operation of biomechanical biofeedback systems largely depends on the parameters of human motion and its analysis algorithms. Biomechanical biofeedback is based on sensing body rotation angles, posture orientation, body translation, and body speed. These parameters are generally calculated from raw data that represent measured physical quantities. For example, the assessment of posture and translations can be performed by accelerometers (gravity, acceleration); body rotations are calculated from the obtained gyroscope data (angular velocity).

Important parameters of human motion should therefore be adequately acquired by the chosen capture system (sensors). Sensors of the motion capture system should have: (a) sufficiently large dynamic ranges for the measured motion quantity, (b) sufficiently high sampling frequency, (c) sufficiently high accuracy and/or precision. The employed processing devices should have sufficient computational power for the chosen analysis algorithms. While this is generally not critical with terminal biofeedback that uses post-processing, it is of the outmost importance with concurrent biofeedback that requires real-time processing. For example, in biofeedback systems with real-time processing all computational operations must be completed within one sampling period. When sampling frequencies are high, the sampling periods are short and this requirement can be quite restricting.

Biofeedback is successful, when the user is appropriately reacting to the given feedback information. Generally that is true when the following conditions and requirements are met: (a) sensing of the body function, parameter, or activity of interest is possible and available with sufficient accuracy and precision; (b) relevant feedback information can be acquired or calculated; (c) appropriate feedback type is used; (d) feedback information is given in the appropriate modality; (e) feedback is available in the proper timing; (f) feedback is understandable to the user; (g) cognitive load of the person in the loop, as the result of feedback information processing, is not too high. The details of these requirements and conditions are studied in Chap. 2.

For example, concurrent biofeedback can only be successfully incorporated when users' reactions are performed in-movement, i.e., inside the time frame of the executed movement pattern. For concurrent in-movement augmented feedback, during the motor task execution, we also need appropriate technical equipment capable of real-time signal acquisition and real-time processing with low communication delay within the feedback loop.

The general requirements of biomechanical biofeedback applications are defined by position and/or orientation tolerance and by the duration of analysis. The typical position errors allowed by biofeedback applications are up to a few centimetres, the typical angular errors are up to a few degrees, and the duration of the motion analysis is typically from a few seconds to a few minutes. For example, in golf biofeedback applications, the required accuracies are $2°$ in orientation and 1 cm in position, and the motion duration is approximately 2 s (Umek et al. 2015). Sensors must exhibit sufficient accuracy, measurement range, and sampling rate to fulfil the above requirements and cover the biofeedback application movement dynamics.

1.2.3 Motor Learning

One of the most common uses of biofeedback is motor learning in sports, recreation, and rehabilitation (Baca et al. 2009; Lauber and Keller 2014). The process of learning new movements is based on repetition (Liebermann et al. 2002). Numerous correct executions are required to adequately learn a certain movement. Biofeedback is successful, if the user is able to either correct errors in a movement or abandon its execution given the appropriate feedback information. Concurrent biofeedback can reduce the frequency of improper movement executions and speed up the process of learning the proper movement pattern. Such movement learning methods are suitable for recreational, professional, and amateur users in the initial stages of the learning process (Liebermann et al. 2002). Initially, this process requires additional learning cycles until the user understands or feels the provided feedback information.

According to sports experts, feedback is the most important variable for motor learning, except the practice itself (Bilodeau et al. 1969). The use of feedback information allows learners to improve their performance, either by correcting an incorrect response or by maintaining a correct response. Feedback is traditionally provided by a trainer or an instructor. The role of the instructor is to stimulate the execution of the correct movements and discourage the execution of incorrect movements. The instructor can be assisted or, in special cases, even replaced by a biofeedback system. During the practice, the natural (inherent) feedback information is provided internally through human sense organs. Augmented feedback is provided by an external source, traditionally by instructors and trainers. Modern technical equipment can help both the learner and the instructor by providing additional, parallel feedback information that is not obtainable by traditional observation methods.

Feedback can be classified based on the timing at which the feedback is provided: (a) feedback that is provided during task execution is known as concurrent feedback, whereas (b) feedback that is provided after the task is referred to as terminal feedback (Sigrist et al. 2013). Depending on the timing of the feedback, the trainee's reaction may be instantaneous or delayed until the next learning period. A biofeedback system works in real time, if it is capable of giving concurrent feedback to a user during the execution of a movement. The concurrent feedback, which is given in real time has been found useful for accelerated motor learning (Baca and Kornfeind 2006; Giggins et al. 2013; Sigrist et al. 2013). One example of a real-time biofeedback system is the application that helps users correct specific golf swing errors (Umek et al. 2015).

1.3 Elements of Biofeedback System

The operation of biofeedback is briefly described in the second paragraph of this chapter. Much more details are given in Chaps. 2–5. This section is a brief introduction to the essential elements needed for its appropriate operation.

1.3.1   Sensing

Biomechanical biofeedback system sensing functionality can be implemented in many different ways. The most traditional are instructor based systems where instructors are using their senses to monitor the performer. Since such biofeedback systems are out of the scope of this book, the focus is on systems where sensing functionality is performed by technical equipment.

The most widespread are various video systems with terminal feedback (Schneider et al. 2015; Chambers et al. 2015). Such systems usually record the exercise or a training episode, which is then replayed and visually analysed shortly after its execution or at any time later. Another group are various motion tracking systems that use reflective markers to track body movement trajectories. Examples of such systems are Vicon (Windolf et al. 2008), OptiTrack (2018), and Qualisys (Josefsson 2002). Such systems may offer real-time functionality, but they are mainly designed to be used in closed and confined spaces with conditions favourable for visual marker tracking. They are also relatively expensive and demand a skilled professional to operate them.

An alternative motion tracking technology is based on inertial sensors. Some advantages of inertial sensor systems are inexpensiveness, accessibility, portability, and ease of use. The drawbacks include possible high inaccuracies of the results when used without error compensation. The most important parameters of movement and of inertial sensor are listed below:

- *Movement dynamics* describes the swiftness of change in a movement, e.g., fast movements in sports and slow movements in rehabilitation biofeedback systems.
- In connection to the above the required biofeedback system *sampling frequency* varies from a few tens to a few hundred Hertz or even higher.
- *Measurement range* defines the minimal and maximal levels of the measured sensor signal (i.e., acceleration signal values of an accelerometer).
- The duration of movement execution defines the width of the *analysis time window* T_w that can vary from less than a second to a few minutes or even hours.
- The *accuracy* of the measured or calculated movement parameters must be sufficiently high for the demands of the application, which vary greatly.
- Measurement *precision* is more important in biofeedback applications than accuracy. For example, when consistently repeating the same movement, measured values must be precise, even if they are not accurate.

1.3.2   Processing

After the motion signals are acquired by the sensors, they must be processed according to the defined signal and data processing algorithms and rules. The result of such processing in the biofeedback system yields feedback information that is forwarded to the user of the system as concurrent or terminal feedback.

Processing functionality of biofeedback systems is provided by great variety of devices; practically all of them are based on digital microchips. Such devices include microcontrollers, microcomputers, smartphones, laptops, personal computers, workstations, and others.

The used processing devices should have sufficient computational power for the chosen processing algorithms. While this is generally not a problem for systems with terminal biofeedback that uses post-processing, it is more critical for systems with concurrent biofeedback that requires real-time processing. Some other requirements and constraints that are connected to the processing are: size, weight, available energy, and autonomy of the processing device of the biofeedback system. A more detailed study of processing and processing devices is given in later chapters of this book.

1.3.3 Feedback

As already explained in Sect. 1.2, feedback is essential for the operation of the biomechanical biofeedback system for accelerated motor learning in sport and rehabilitation.

In biomechanical biofeedback systems two basic feedback types exist in respect to its source; intrinsic (natural) and extrinsic (augmented). In contrast to natural or intrinsic biofeedback, which is based on proprioception, augmented or extrinsic biofeedback relies on information from an outside source; i.e. from instructors or from artificial devices (Alahakone and Senanayake 2009; Crowell et al. 2010; Franco et al. 2013; Huang et al. 2006).

Similarly to sensing and processing, feedback can be done either by instructors or by technical devices. This book concentrates on biomechanical biofeedback systems with augmented feedback provided by technical devices.

Feedback Modality

A biofeedback system can communicate with the human nervous system through various modalities, i.e., visual, auditory or tactile. User's sensing ability and sense occupancy during the performed task condition the choice of modality. For example, in many cases visual feedback can interfere with the user's intensive visual preoccupation with the performed task and can therefore be rather distracting. Another drawback of the visual modality is its high cognitive load. The tactile feedback modality is becoming an increasingly more appropriate alternative for many applications in sports and medical rehabilitation (Alahakone and Senanayake 2009; Crowell et al. 2010; Kirby 2009; Lieberman and Breazeal 2007). In many cases the best choice is auditory feedback, because auditory sensing is mostly unengaged during exercise. The other reasons for this choice can be more pragmatic; namely, auditory sensing can be readily implemented, it has low cognitive load, and the same audio channel can be used for both biofeedback and application control. The advantages of using the auditory feedback channel for motor learning are analysed and presented in (Dozza et al. 2011; Eriksson et al. 2011; Schaffert et al. 2010; Vogt et al. 2010).

1.3.4   System Example

One example of a biomechanical biofeedback application in sports that can illustrate some of the above discussed properties and requirements, is an application for golf swing error detection. Golf swing is a static exercise, but the body movements can be rather fast (high dynamic). Consequently, the sensor sampling frequency must be high enough. Depending on the sensor placement, the accelerometer must cover different measurement ranges. For example, when attached to the lower part of the golf club, accelerations are much higher than when attached to the wrist or the arm of the player. The analysis time frame is between 1.5 and 3 s, depending on the player. The required accuracies are between 2° and 3° for posture and angular rotation. The required measurement consistency is in the range of 2°. The developed experimental real-time biomechanical biofeedback application uses smartphone sensors for sensing, laptop for processing and headphones for auditory feedback (Umek et al. 2015).

1.4   Technology

Technology used in biofeedback systems in sport and rehabilitation is developing very fast; recent day technology possesses properties and functionalities only wished-for a few years ago. For example, in the past the motion of gymnasts could only be analysed in certain detail through video recordings, while at present gymnasts can wear a suit with motion sensors (The Xsens 2018) that records their moves. Based on the athlete's kinematic model such systems can give a detailed analysis of their motion in three-dimensional space. Similar examples could be found for other sports.

  The popularity of wearable devices and smartphones has lowered the cost of various sensors used for monitoring users' activity and tracking their motion. From simple applications that measure the calories burnt and number of steps made, sensors are now moving to sports equipment (Lightman 2016). Accelerometers, gyroscopes, magnetometers, and other sensors are being integrated into golf clubs, basketballs, tennis rackets, baseball bats, and other sport equipment, making it smart sport equipment. Wearable devices and smart sport equipment are slowly but surely bringing the Internet of Things (IoT) into sport, recreation, and our daily life (Ebling 2016). In general, science and technology are increasingly used to augment sport training and exercise (Kunze et al. 2017). According to the above said the ubiquitous computing concept that presumes the need for computing anytime and anywhere, is becoming indispensable (Baca et al. 2009).

1.4.1 Quantification

Modern society is developing in many different directions. Individuals are becoming more self-centred and self-conscious about themselves, about their health, and about their well-being. Physical activity is becoming an increasingly important aspect of our lives. It is a necessary and a required ingredient of a healthy life and there is no doubt that it contributes to our wellbeing.

By incorporating data acquisition technology into our daily life, we can measure and quantify our consumption of food, air, and liquids; physical and mental performance; physical and physiological states. We can even acquire biometric information. Self-monitoring and self-sensing systems can combine wearable sensors with wearable computing functionalities. One subset of self-sensing systems based on wearable sensors and wearable computing includes motion tracking and movement recognition systems and their applications.

Comparison, quantification, and competition are in the human nature and sport is a perfect medium and technology provides the means for all of the abovementioned. The following sections briefly introduce technologies that are used in biomechanical biofeedback systems.

1.4.2 Sensors

Sensors and motion caption systems are the essential components of the biomechanical biofeedback system with augmented feedback. The system works with one or multiple sensors or sensor devices that are usually attached to the user's body or they are somewhere in user's vicinity. Sensors are the source of signals and data used by the processing device.

Motion Caption and Motion Tracking Systems

Motion capture systems usually include multiple sensors of the same type. The most common are camera based systems and inertial sensor based systems (Liebermann et al. 2002).

Motion tracking can be performed at different scales—from low-precision tracking and navigation in closed buildings (Hardegger et al. 2015) with the required accuracies in metres to the submillimetre scale in the high-precision tracking of the fine movement of a speaker's lips and jaw (Feng and Max 2014). Various human body motion tracking applications require accuracy in the range of centimetres and fall somewhere between these extremities.

In many sports disciplines camera based systems that include recording and tracking are classical methods for providing augmented feedback information for post analysis and terminal feedback. Camera based systems can be divided into two main subcategories: (a) video based systems and (b) marker based systems (Qualisys 2018; Vicon 2018). The former directly process the video stream captured at various light wavelengths, the later use passive or active markers for determining their three dimensional position in space and time.

In recent years, with the rapid development of lightweight sensors and sensor devices, inertial sensor based motion capture systems are becoming more popular (iSen 2018; Tracklab 2018; Xsens 2018). It should be emphasized that inertial sensor based motion tracking systems are generally mobile and in contrary to camera based systems have no limitation in space coverage. Modern inertial sensors are miniature low-power chips integrated into wearable sensor devices or smart equipment. Wearable and integrated sensors should measure the monitored quantities, but they should not interfere with the activity itself. Therefore, such sensors must be lightweight and small-size and they should not physically obstruct the activity. The last requirement also implicitly defines the wireless mode of communication between sensors and processing devices.

Some advantages of inertial sensor systems are inexpensiveness, accessibility, portability, and ease of use. The drawbacks include possible high inaccuracies of the results when used without error compensation.

MEMS Inertial Sensors

Today's inertial sensors predominantly fall into the group of micro electromechanical systems (MEMS). MEMS inertial sensors are portable, miniature, lightweight, inexpensive, and low power. Because of their properties, they are often the first choice for integration into wearable devices used in motion tracking systems. Such sensors have also been adopted in various electronic products, e.g., smartphones, tablets, game consoles, and wearable sensor devices. MEMS gyroscopes and accelerometers offer lower performance compared with professional navigational sensors (Grewal 2010). The precision of MEMS gyroscopes and accelerometers is primarily affected by their biases, which induce errors in their derived angular and spatial positions. However, the imprecision of measurements from accelerometers and gyroscopes is not particularly critical in biofeedback applications, when used in short-time signal analysis (Kos et al. 2016; Umek et al. 2015; Umek and Kos 2016).

To investigate the limitations of MEMS inertial sensors for use in biofeedback systems, the general properties and demands of biomechanical biofeedback applications should be defined. In biomechanical biofeedback applications, inertial sensors are used to detect and possibly track body movements. Various parameters can be used for the evaluation of inertial sensors for particular biomechanical biofeedback applications; those parameters are listed in Sect. 1.3.1.

Inertial sensors, particularly accelerometers and gyroscopes, are used in many applications in sports, recreation, rehabilitation, and wellbeing. One particular use of accelerometers and gyroscopes is body motion tracking in biomechanical biofeedback systems. For example, a rehabilitation system with three inertial sensor units was used for post-processing for exercise performance assessment in Giggins et al. (2014). In sports training, the movement dynamics are usually greater than those in healthcare and rehabilitation, which limits the possible usage of MEMS inertial sensors for biomechanical movement biofeedback. In Crowell et al. (2010), accelerometers attached to a runner's leg were used to drive real-time visual feedback with the intent of reducing tibial stress. The paper from Schneider et al. (2015) reviews the use of sensors in the learning application domain and analyses 82 sensor-based prototypes with respect to their learning support. Another paper from Chambers et al. (2015) provides a review of the current ability of MEMS sensors to detect sport-specific movements and demonstrates that commercially available micro sensors are capable of quantifying sporting demands that other monitoring technologies may not detect. Many more similar examples can be easily found in research works in the fields of biomechanical biofeedback systems.

1.4.3 Devices

Biomechanical biofeedback systems with augmented feedback include one or several devices that perform any combination of sensing, processing, and feedback. Practically all such devices are digital electronic devices based on microchips.

Depending on their function, such devices may have very different properties. In terms of size and weight, devices range from miniature wearable devices including MEMS sensors, to large heavy motion capture systems, mounted in the fixed space. In terms of energy source they can be autonomous battery powered devices or supercomputers using large amounts of energy. Here we give a brief presentation of the most representative groups of devices characteristic to biofeedback systems with the emphasis on the devices using MEMS inertial sensors. More detailed explanations are given in Chap. 6.

Wearables

More devices are being made wearable due to advancements in miniaturization. Wearables are among the enablers of ubiquitous and mobile computing, making the technology pervasive by interweaving it into our personal lives. Various quantifying systems that are used for self-sensing and self-monitoring combine wearable sensors and devices, data acquisition techniques, and wearable computing (Chambers et al. 2015; Hardegger et al. 2015; Varkey et al. 2012; Wong et al. 2015).

Wearables come in a great variety, from smart devices, such as smartphones or smart watches, to small accessories, such as smart eyewear or ear-buds. They are worn on the body as accessories, attachments or implants. A comprehensive survey of wearable devices and their challenges can be found in Seneviratne et al. (2017); their technological challenges are also studied in Williamson et al (2015).

Wearable devices can play an important role in increasing the QoL (Park and Jayaraman 2003), which is one of the primary aims of individuals and society as a whole. For example, the pervasiveness and the personalization of healthcare go hand in hand with wearable devices (Zhang et al. 2018). The evolution of pervasive healthcare based on wearables, from wearable sensors for activity detection to smart implants, is studied in Andreu-Perez et al. (2015). The authors assert that pervasive healthcare is moving towards preventative, predictive, personalized, and participatory functions. Smart wearables will enable not only the monitoring of common physiological parameters, such as body temperature, blood pressure, and heart rate but also environmental parameters, geographical location, physical activity detection and recognition, and other measurable parameters. Such variety of data will enable advanced healthcare services, such as behaviour profiling, for promoting healthy living with objective feedback (Ali 2013) and ontology-driven interactive healthcare supported by real-time user context information (Kim et al. 2014).

Smartphones

According to Zenith (2018) the penetration of smartphones in 52 countries, representing 65% of the world's population, will reach 66% in 2018. In the most advanced countries this number is even higher, reaching between 80 and 90%. Consequently, smartphone applications are penetrating the fields of sports, health monitoring, human behavioural monitoring, and social networking (Feese et al. 2014; Gravenhorst et al. 2014; Xia et al. 2013; Zhang et al. 2014). These applications utilise the built-in sensors and functionalities of smartphones, and some also employ additional specialised and wirelessly connected wearable devices.

All present-day high-end smartphones are equipped with MEMS accelerometers and gyroscopes, many also with magnetometers and other sensors that can be used in biofeedback applications. Smartphone MEMS sensors are sufficiently precise for use in real-time biomechanical movement biofeedback applications if the analysis time window is not excessively large and if the movement dynamics fall within the MEMS dynamic range (Umek et al. 2015).

Various applications of smartphone MEMS sensors in biomechanical biofeedback systems can be found in the fields of healthcare, rehabilitation, and sports. In Dai et al. (2010), accelerometers in smartphones are used to detect and immediately communicate falling events of patients under care. In Casamassima et al. (2014), a wearable system for gait training was demonstrated that provided individuals with Parkinson's disease with vocal feedback by encouraging them to maintain or correct their walking behaviour. Pernek et al. (2013) used smartphone accelerometers for the detection of activity types and to facilitate repetition analysis, of which the results were made available in statistical form following training. Smartphone motion sensors were applied in Wei et al. (2014) to detect and evaluate the performances of dance students compared with those of dance trainers. The movement analysis results were made available to the user immediately following the dance episodes.

Using smartphones in biofeedback systems has many advantages. Practically all relatively new smartphones are laden with a number of different sensors: accelerometers, gyroscopes, microphones, GPS, cameras, magnetometers, etc. In addition to that, smartphones also include feedback devices: loudspeakers, screen, and vibration device. And because smartphones have a considerable amount of processing power, they can be used for the implementation of standalone mobile applications. That means that the smartphone acquires activity parameters, calculates results, and gives feedback through one of its feedback devices. Another advantage is the synchronization of all sensor signals taken from the same smartphone. Less demanding biofeedback applications can be implemented entirely on the smartphone. Use of smartphones could also have some disadvantages. The most notable are their size and weight. When used as motion tracking sensors, smartphones (a) cannot be physically attached to certain parts of the body and (b) could interfere with the movement being executed due to their weight and size. Size limits the choice of body attachment points, and weight is the limiting factor in large dynamic movements. Challenging environments, such as water, and operating-system restraints on the sensor performance parameters, such as limiting the sampling frequency far below the MEMS sensor capabilities, also narrow the use of smartphones.

Gadgets

In recent years a number of inexpensive wearable monitoring devices and gadgets aiming for the activity tracking have been introduced to the market. Gadgets, such as wrist bands, smart-watches or pendants, give statistical parameters and count events of a particular physical activity. For example, they count the number of steps or stairs made during the day, they can detect falls, or estimate sleep quality and stress levels. Such gadgets usually acquire movements or physiological processes of the user with low frequency and low precision, what is at the end good enough for their intended use. At the other end of sport technology are complex and expensive systems that simultaneously gather and process large amounts of data. For example, a system for a real-time tracking of a football match and the analysis of training (von der Grün et al. 2011). The majority of applications in sport based on technology lie somewhere between both abovementioned groups.

This book studies the possible use of embedded devices in rehabilitation and sports. We try to make a leap from toys and gadgets that mostly offer approximate, statistically based, activity and biometric measurements, to devices, tools, and applications that would offer precise and timely information for motor learning or training improvement.

Smart Sport Equipment

Professional, often also amateur and recreational sports are highly competitive. Gaining even a small advantage may result in winning, especially in professional sport. The use of science and technology offers various ways of getting such advantage(s) (Giblin et al. 2016; Lightman 2016). One field of research is the use of smart sport equipment (Umek and Kos 2018).

In sport, sensor technology has been predominantly used for motion tracking and analysis with wearable motion sensors attached to the athlete's body. In recent years biofeedback applications are appearing in various sports, and sensors are being built into sport equipment as well. Biofeedback systems are important in motor learning in sports where users employ the feedback information to influence their execution performance (Giblin et al. 2016).

In many sports the equipment is an inseparable element of the action. Athletes use sport equipment as a tool or medium through which their energy and actions are transferred into a desired result. The equipment can be as simple as a baseball bat, or it can be as complex as a Formula 1 car. Measuring and quantifying the actions of the athlete and the response of the sport equipment is expected to prove beneficial to athlete performance improvement. Various sensors can be integrated into smart sport equipment; they should measure the relevant and desired quantities, but they should not interfere with the sport activity itself. Therefore sensors must be lightweight and small-size, they should not alter the properties and functionality of the equipment, and they should not physically obstruct the activity. For the detailed quantification of sport activities a number of physical and physiological quantities should be measured simultaneously; from heart rate and body temperature, to exerted forces, material bends, accelerations, and rotation speeds, among others.

For complex sport equipment the technology has always played a major role in getting the competitive advantage over the opponents. For example, technologically superior bob sledge can eventually win over the technologically inferior one, even if its team is not as good. The technology is now making its way also to the simple sport equipment. Manufacturers of sport equipment have already put to market several examples of smart sport equipment, such as smart tennis racket, smart basketball, smart running shoes, and others (Lightman 2016). While simple sport equipment might not require complex technology, it might be difficult or even impossible to design because of its size and weight restrictions, its possible violent use, or for any other reason.

1.4.4 Communication

Communication channels enable the transmission of signals and data between the biofeedback system devices. Although wireless communication technologies are most commonly used, wired technologies are also used in practice. In the case of implants, the human body is used as the propagation channel (Cavallari et al. 2014). With local processing, where all the devices are close to each other, wired and wireless technologies can be used. With remote processing, where devices are distributed over a larger physical area, only wireless technologies are practical.

The choice of the particular communication channel heavily depends also on the type and the dynamics of the sport's activity being monitored. For example, static sports or sports with very low dynamics may allow the use of wired sensors, while high dynamic sports with a lot of movement would not. Concerning the requirement for minimum obstruction of the user the most appropriate are systems with wireless communication.

The most demanding are biomechanical biofeedback systems with concurrent feedback, where communication must be performed in real time, that is, with as little delay as possible.

Wireless Technologies

Wireless communication can represent an obstacle in implementation of biomechanical biofeedback systems. For example, when trying to implement a real-time system for a high dynamic sport, a number of serious obstacles can be identified in sensor and processing devices: sensors may have insufficient dynamic range, sampling frequency and/or accuracy, processing devices may have insufficient processing power or consume too much energy, etc.

An important part of the problem are obstacles that are most commonly found in available wireless communication technologies; some of them are described in Pyattaev et al. (2015). The limitations of wireless technologies are represented by interconnected transmission parameters: range, power, and bit rate. For example, actions of sport activity are taking place in an area that cannot be covered by the chosen wireless technology or the amount of streaming data is too large for the bit rates of the available wireless technology. The physical communication channel limitations drive the search for the best balance of achievable levels of bit rate and range against the available or prescribed transmission power. Wireless technologies are presented and thoroughly studied in Chap. 6.

Internet of Things

A technology inherent to the concept of pervasive computing is the Internet of Things (IoT). The pervasive data collection requires a great number of different network-connected devices (things). This is making the IoT a technology of choice for the concept of pervasive computing, that nicely fits into the concept of biofeedback systems. Opportunities and challenges for the IoT have been first recognized in healthcare and most of them see the highest potential in wearable devices (wearables) as a part of healthcare IoT (Fernandez and Pallis 2014; Metcalf et al. 2016). Hiremath et al. (2014) even go a step further by defining the new concept of a Wearable Internet of Things for person-cantered healthcare. The concept of IoT, as described in the relation to healthcare, is equally applicable to biofeedback systems.

According to Tronconi (2013) the key enablers of the Internet of Things are MEMS sensors. In most cases, smart devices (things) include sensors, microcontrollers, and low power wireless communication. Used together they pave the road to smart system; among others in the fields of health, wellness, recreation, rehabilitation, and sports.

We are particularly interested in the role of IoT in human wellbeing, which includes, but is not limited to healthcare, rehabilitation, recreation, and sports. Sport and recreation are identified as some of the most rapidly growing areas of personal and consumer Internet of Things applications (Lightman 2016). Smart devices or smart wearables such as wrist band activity trackers, heart rate monitors, motion tracking devices, and others are penetrating our daily life; they are readily available, affordable, and fast growing in numbers, and spatial density. The practical use of such devices heavily depends on the development of wireless and sensor technology, corresponding applications, and a close cooperation with sports experts.

Example—Smart Sport Equipment and IoT Applications in Sport

Smart sport equipment is one group of smart IoT devices that are used in the field of sport and recreation. Smart sport equipment includes at least one sensor and is communicating to another device for sensor data processing and analysis. The communication is wired or wireless, it can be local or through one or more networks (Saintoyant and Mahonen 2006). The processing can be performed locally, by the

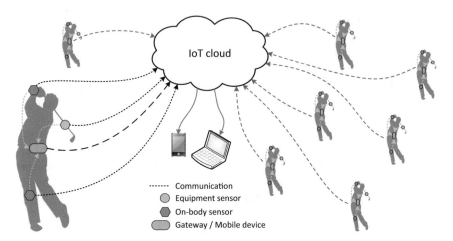

Fig. 1.1 Simplified architecture of IoT sport applications using smart sport equipment and wearables with integrated sensors. Smart sport equipment and wearables send sensor data to the IoT cloud directly or through the gateway. Sensor data processing can be performed locally by the mobile device or in the cloud. Results can be checked by any connected device (Kos et al. 2018)

(mobile) device located near or inside the smart sport equipment, or remotely, for example in the cloud.

Smart sport equipment connected to internet forms one group of sport IoT devices. Another group of sport IoT devices are wearable devices attached to the user's body. Both groups of the abovementioned devices can be used in various mobile and IoT sport applications. A simplified architecture of IoT cloud for sport applications is shown in Fig. 1.1. In general, every IoT sport application implements processes of data acquisition, data transmission, data processing, data storage, and presentation of data and/or processing results. While data acquisition is always at the beginning and data presentation is at the end of the process chain, the rest of the processes are not necessarily always in the same order. On the basis of the above said, three basic groups of IoT sport applications can be identified:

- In the classical IoT application sport IoT devices send their data directly to the cloud. In Fig. 1.1 this is shown with black dotted lines. Data is processed and stored in the cloud. The downsides of such application are possible problems with the synchronization of data from independent sport IoT devices, which is caused by different latencies of their independent transmission channels.
- To avoid the abovementioned problems, sport IoT devices first send their data to the gateway as shown by green dotted lines in Fig. 1.1. The data is synchronized and processed locally. The results are sent to the cloud for storage and presentation (see black dashed line in Fig. 1.1). The downside of such applications is the possible need for large processing power, which might not be available locally.
- The hybrid solution is to collect and synchronize IoT devices data locally. The merged raw data stream is then sent to the cloud for processing, storage and

presentation. The raw data path is presented by green dotted lines and black dashed line in Fig. 1.1.

Data acquisition is done by the smart sport equipment sensors and on-body sensors in wearable devices; the presentation of results is performed by connected devices, as shown by blue lines in Fig. 1.1. Many different sport IoT applications belonging to the one of the above basic groups are envisioned to be active at the same time. In Fig. 1.1 they are represented by the minimized silhouettes connected to the cloud by red dashed lines.

1.5 Vision

Our vision is to design and implement biofeedback systems and applications that would be able to satisfy a wide range of possible uses and that would also support the use of smart equipment. For example, an application for running would be implemented on a smartphone. It would be able to give real time feedback to the user about some basic running parameters, such as left and right leg period balance and similar. Users of this application would most probably be able to improve their running technique, if given some advice by an expert (instructor).

Another viable example is a biofeedback system that would give real time information about user's performance to the instructor only. The instructor would then decide if immediate feedback to the user is necessary or not. Such system could be also used for later more detailed analysis and terminal feedback to the user and/or to the instructor.

Some of the future efforts lie also in machine learning, which is an effective technique of artificial intelligence. Our vision is to designing system that would collect motion data, learn from them based on machine learning algorithms and methods, and then use the acquired knowledge to perform real-time biofeedback to the user. With such methods, we can receive professional advices at any time without constantly asking instructors for help. Let us illustrate this vision on a golf example. Such a system would collect a number of golf swings from the golf player. In the first phase it would learn which swings are performed correctly and which are not. In the second phase the system would operate as a real time biofeedback system that would monitor each swing execution. The system would alert the player about incorrectly started swings that in great majority also end incorrectly, what results in unwanted ball flight. If the player would abort the execution of incorrectly started swing before the impact, such swing would not be memorized. Consequently, that would result in faster motor learning and at the end also in better score in the field.

There are many more similar examples from different fields of sport and rehabilitation. Some of the implemented ideas are presented in Chap. 7.

Scope and Organization of the Book

The scopes of this book are biomechanical biofeedback systems and applications with augmented feedback that is provided by technical equipment. That means that such systems and applications use technology to implement at least one of the functions of sensing, processing, and feedback. Some more attention is given to the subgroup of real-time biofeedback systems that provide concurrent feedback.

The book is organized in chapters that present the topic of biomechanical biofeedback systems and applications in an organized and structured manner.

- Chapter 2 *Biomechanical Biofeedback* presents its background, basics, and related topics that are essential for better understanding of the material given in the subsequent chapters.
- Chapter 3 *Biofeedback System* studies the technical implementation of biomechanical biofeedback systems. It describes its operation and its technical building blocks.
- Chapter 4 *Biofeedback System Architectures* build on previous chapters with in-depth examination of possible architectures of biomechanical biofeedback systems in terms of its properties, constraints, and intended functionalities. The chapter is concluded with the classification and comparison section, which presents a classification graph and comparison of the architectures.
- Chapter 5 *Biofeedback Systems in Sport and Rehabilitation* studies the main challenges of biofeedback systems in sport and rehabilitation. The stress is on real-time systems with concurrent feedback. The chapter presents aspects of the properties of feedback loop elements that are important for applications in sport and rehabilitation. It also identifies some possible obstacles for implementation of such systems in sport.
- Chapter 6 *Performance Limitations of Biofeedback System Technologies* discusses properties and limitations of the selected technologies crucial for the appropriate and timely operation of the biofeedback system. The emphasis is on inertial sensors and wireless communication technologies.
- Chap. 7 *Applications* presents and discusses five examples of biofeedback applications developed by the authors of this book: Golf swing trainer, Smart golf club, Smart ski, water sports, and swimming rehabilitation. Each of them is thoroughly explained in terms of its objectives and functionalities, system architecture and setup, theoretical and research background, results and future development plans.

References

Alahakone AU, Senanayake SA (2009) A real time vibrotactile biofeedback system for improving lower extremity kinematic motion during sports training. In: International conference of soft computing and pattern recognition, 2009. SOCPAR'09. IEEE, pp 610–615, Dec 2009

Ali SMR (2013) Behaviour profiling using wearable sensors for pervasive healthcare

Andersen LB, Harro M, Sardinha LB, Froberg K, Ekelund U, Brage S, Anderssen SA (2006) Physical activity and clustered cardiovascular risk in children: a cross-sectional study (The European Youth Heart Study). Lancet 368(9532):299–304

Andreu-Perez J, Leff DR, Ip HM, Yang GZ (2015) From wearable sensors to smart implants—toward pervasive and personalized healthcare. IEEE Trans Biomed Eng 62(12):2750–2762

Baca A, Kornfeind P (2006) Rapid feedback systems for elite sports training. IEEE Pervasive Comput 5(4):70–76

Baca A, Dabnichki P, Heller M, Kornfeind P (2009) Ubiquitous computing in sports: a review and analysis. J Sports Sci 27(12):1335–1346

Becker BE (2009) Aquatic therapy: scientific foundations and clinical rehabilitation applications. PM&R 1(9):859–872

Bilodeau EA, Bilodeau IM, Alluisi EA (1969) Principles of skill acquisition. Academic Press

Camomilla V, Bergamini E, Fantozzi S, Vannozzi G (2016). In-field use of wearable magneto-inertial sensors for sports performance evaluation. In: ISBS-conference proceedings archive, vol 33, No 1, May 2016

Casamassima F, Ferrari A, Milosevic B, Ginis P, Farella E, Rocchi L (2014) A wearable system for gait training in subjects with Parkinson's disease. Sensors 14(4):6229–6246

Cavallari R, Martelli F, Rosini R, Buratti C, Verdone R (2014) A survey on wireless body area networks: technologies and design challenges. IEEE Commun Surv Tutor 16(3):1635–1657

Chambers R, Gabbett TJ, Cole MH, Beard A (2015) The use of wearable microsensors to quantify sport-specific movements. Sports Med 45(7):1065–1081

Crowell HP, Milner CE, Hamill J, Davis IS (2010) Reducing impact loading during running with the use of real-time visual feedback. J Orthop Sports Phys Ther 40(4):206–213

Dai J, Bai X, Yang Z, Shen Z, Xuan D (2010) Mobile phone-based pervasive fall detection. Pers Ubiquit Comput 14(7):633–643

Dozza M, Chiari L, Peterka RJ, Wall C, Horak FB (2011) What is the most effective type of audio-biofeedback for postural motor learning? Gait Posture 34(3):313–319

Ebling MR (2016) IoT: from sports to fashion and everything in-between. IEEE Pervasive Comput 4:2–4

Eriksson M, Halvorsen KA, Gullstrand L (2011) Immediate effect of visual and auditory feedback to control the running mechanics of well-trained athletes. J Sports Sci 29(3):253–262

Feese S, Burscher MJ, Jonas K, Tröster G (2014) Sensing spatial and temporal coordination in teams using the smartphone. Hum-Centric Comput Inf Sci 4(1):1–18 (Springer, Berlin, Heidelberg)

Feng Y, Max L (2014) Accuracy and precision of a custom camera-based system for 2-D and 3-D motion tracking during speech and nonspeech motor tasks. J Speech, Lang Hear Res 57(2):426–438

Fernandez F, Pallis GC (2014) Opportunities and challenges of the Internet of Things for healthcare: systems engineering perspective. In: 2014 EAI 4th international conference on wireless mobile communication and healthcare (Mobihealth). IEEE, pp. 263–266, Nov 2014

Franco C, Fleury A, Guméry PY, Diot B, Demongeot J, Vuillerme N (2013) iBalance-ABF: a smartphone-based audio-biofeedback balance system. IEEE Trans Biomed Eng 60(1):211–215

Giblin G, Tor E, Parrington L (2016) The impact of technology on elite sports performance. Sensoria: J Mind, Brain Cult 12(2)

Giggins OM, Persson UM, Caulfield B (2013) Biofeedback in rehabilitation. J Neuroeng Rehabil 10(1):60

Giggins OM, Sweeney KT, Caulfield B (2014) Rehabilitation exercise assessment using inertial sensors: a cross-sectional analytical study. J Neuroeng Rehabil 11(1):158

Gravenhorst F, Muaremi A, Bardram J, Grünerbl A, Mayora O, Wurzer G, …, Tröster G (2014) Mobile phones as medical devices in mental disorder treatment: an overview. Pers Ubiquit Comput 1–19

Grewal M, Andrews A (2010) How good is your gyro [ask the experts]. IEEE Control Syst 30(1):12–86

Gruwsved Å, Söderback I, Fernholm C (1996) Evaluation of a vocational training programme in primary health care rehabilitation: a case study. Work 7(1):47–61

Hanson S, Jones A (2015) Is there evidence that walking groups have health benefits? A systematic review and meta-analysis. Br J Sports Med 49(11):710–715

Hardegger M, Roggen D, Tröster G (2015) 3D ActionSLAM: wearable person tracking in multi-floor environments. Pers Ubiquit Comput 19(1):123–141

Hiremath S, Yang G, Mankodiya K (2014) Wearable Internet of Things: concept, architectural components and promises for person-centered healthcare. In: 2014 EAI 4th international conference on wireless mobile communication and healthcare (Mobihealth), Nov 2014

Huang H, Wolf SL, He J (2006) Recent developments in biofeedback for neuromotor rehabilitation. J Neuroeng Rehabil 3(1):11

iSen, Inertial motion capture (2018) https://www.stt-systems.com/products/inertial-motion-captur e/isen/. Accessed 23 June 2018

Josefsson T (2002) U.S. Patent No. 6,437,820. U.S. Patent and Trademark Office, Washington, DC

Kim J, Kim J, Lee D, Chung KY (2014) Ontology driven interactive healthcare with wearable sensors. Multimed Tools Appl 71(2):827–841

Kirby R (2009) Development of a real-time performance measurement and feedback system for alpine skiers. Sports Technol 2(1–2):43–52

Kos A, Umek A (2018) Wearable sensor devices for prevention and rehabilitation in healthcare: swimming exercise with real-time therapist feedback. IEEE Internet Things J. https://doi.org/10. 1109/jiot.2018.2850664

Kos A, Milutinović V, Umek A (2018) Challenges in wireless communication for connected sensors and wearable devices used in sport biofeedback applications. Future Gener Comput Syst

Kos A, Tomažič S, Umek A (2016) Suitability of smartphone inertial sensors for real-time biofeedback applications. Sensors 16(3):301

Kunze K, Minamizawa K, Lukosch S, Inami M, Rekimoto J (2017) Superhuman sports: applying human augmentation to physical exercise. IEEE Pervasive Comput 16(2):14–17

Lauber B, Keller M (2014) Improving motor performance: Selected aspects of augmented feedback in exercise and health. Eur J Sport Sci 14(1):36–43

Lieberman J, Breazeal C (2007) Development of a wearable vibrotactile feedback suit for accelerated human motor learning. In: 2007 IEEE international conference on robotics and automation. IEEE, pp 4001–4006, Apr 2007

Liebermann DG, Katz L, Hughes MD, Bartlett RM, McClements J, Franks IM (2002) Advances in the application of information technology to sport performance. J Sports Sci 20(10):755–769

Lightman K (2016) Silicon gets sporty. IEEE Spectr 53(3):48–53

Lohne-Seiler H, Hansen BH, Kolle E, Anderssen SA (2014) Accelerometer-determined physical activity and self-reported health in a population of older adults (65–85 years): a cross-sectional study. BMC Public Heal 14(1):284

McGrath MJ, Scanaill CN (2013) Wellness, fitness, and lifestyle sensing applications. In: Sensor technologies. Apress, Berkeley, CA, pp 217–248

Metcalf D, Milliard ST, Gomez M, Schwartz M (2016) Wearables and the internet of things for health: wearable, interconnected devices promise more efficient and comprehensive health care. IEEE Pulse 7(5):35–39

Nagle EF, Sanders ME, Franklin BA (2017) Aquatic high intensity interval training for cardiometabolic health: benefits and training design. Am J Lifestyle Med 11(1):64–76

Neiva HP, Marques MC, Travassos BF, Marinho DA (2017) Wearable Technology and aquatic activities: a review. Motricidade 13(1):219

OptiTrack (2018) http://optitrack.com/hardware/. Accessed 30 June 2018

Park S, Jayaraman S (2003) Enhancing the quality of life through wearable technology. IEEE Eng Med Biol Mag 22(3):41–48

Pernek I, Hummel KA, Kokol P (2013) Exercise repetition detection for resistance training based on smartphones. Pers Ubiquit Comput 17(4):771–782

Prins J, Cutner D (1999) Aquatic therapy in the rehabilitation of athletic injuries. Clin Sports Med 18(2):447–461

Pyattaev A, Johnsson K, Andreev S, Koucheryavy Y (2015) Communication challenges in high-density deployments of wearable wireless devices. IEEE Wirel Commun 22(1):12–18

Qualisys, Motion Capture System (2018) http://www.qualisys.com. Accessed 10 June 2018

Ruiz JR, Ortega FB, Rizzo NS, Villa I, Hurtig-Wennlöf A, Oja L, Sjöström M (2007) High cardiovascular fitness is associated with low metabolic risk score in children: the European Youth Heart Study. Pediatr Res 61(3):350–355

Saintoyant PY, Mahonen P (2006) U.S. Patent Application No. 11/030,217

Schaffert N, Mattes K, Effenberg AO (2010) A sound design for acoustic feedback in elite sports. In: Auditory display. Springer, Berlin, Heidelberg, pp 143–165

Schneider J, Börner D, Van Rosmalen P, Specht M (2015) Augmenting the senses: a review on sensor-based learning support. Sensors 15(2):4097–4133

Seneviratne S, Hu Y, Nguyen T, Lan G, Khalifa S, Thilakarathna K, …, Seneviratne A (2017) A survey of wearable devices and challenges. IEEE Commun Surv Tutor 19(4):2573–2620

Sigrist R, Rauter G, Riener R, Wolf P (2013) Augmented visual, auditory, haptic, and multimodal feedback in motor learning: a review. Psychon Bull Rev 20(1):21–53

Silva ASM (2014) Wearable sensors systems for human motion analysis: sports and rehabilitation

Singh R, Stringer H, Drew T, Evans C, Jones RS (2015) Swimming breaststroke after total hip replacement; are we sending the correct message. J Arthritis 4(147):2

The Xsens wearable motion capture solutions (2018) https://www.xsens.com/products/xsens-mvn/. Accessed 10 June 2018

Tomažič S (2006) Quality of life: a challenge for engineers?. In: International conference on advances in the internet, processing, systems and interdisciplinary research, Montreal, New York, Boston

Tracklab, Inertial Motion Capture Systems (2018) https://tracklab.com.au/inertial-motion-capture-systems/. Accessed 23 June 2018

Tronconi M (2013) MEMS and Sensors are the key enablers of Internet of Things. STMicroelectronics: Geneva, Switzerland

Umek A, Kos A (2016) Validation of smartphone gyroscopes for mobile biofeedback applications. Pers Ubiquit Comput 20(5):657–666

Umek A, Kos A (2018) Smart equipment design challenges for real time feedback support in sport. Facta Universitatis, Series: Mechanical Engineering

Umek A, Tomažič S, Kos A (2015) Wearable training system with real-time biofeedback and gesture user interface. Pers Ubiquit Comput 19(7):989–998

Varkey JP, Pompili D, Walls TA (2012) Human motion recognition using a wireless sensor-based wearable system. Pers Ubiquit Comput 16(7):897–910

Vicon, Camera Systems (2018) https://www.vicon.com/products/camera-systems. Accessed 23 June 2018

Vogt K, Pirró D, Kobenz I, Höldrich R, Eckel G (2010) PhysioSonic-evaluated movement sonification as auditory feedback in physiotherapy. In: Auditory display. Springer, Berlin, Heidelberg, pp 103–120

von der Grün T, Franke N, Wolf D Witt N, Eidloth A (2011) A real-time tracking system for football match and training analysis. In: Microelectronic systems. Springer, Berlin, Heidelberg, pp 199–212

Wei Y, Yan H, Bie R, Wang S, Sun L (2014) Performance monitoring and evaluation in dance teaching with mobile sensing technology. Pers Ubiquit Comput 18(8):1929–1939

Williamson J, Liu Q, Lu F, Mohrman W, Li K, Dick R, Shang L (2015) Data sensing and analysis: challenges for wearables. In: 2015 20th Asia and South Pacific design automation conference (ASP-DAC). IEEE, pp 136–141, Jan 2015

Windolf M, Götzen N, Morlock M (2008) Systematic accuracy and precision analysis of video motion capturing systems—exemplified on the Vicon-460 system. J Biomech 41(12):2776–2780

Wong C, Zhang ZQ, Lo B, Yang GZ (2015) Wearable sensing for solid biomechanics: a review. IEEE Sens J 15(5):2747–2760

Xia F, Hsu CH, Liu X, Liu H, Ding F, Zhang W (2013) The power of smartphones. Multimed Syst 21(1):87–101

Yilmaz I, Yanardag M, Birkan B, Bumin G (2004) Effects of swimming training on physical fitness and water orientation in autism. Pediatr Int 46(5):624–626

Zenith, Smartphone penetration (2018) https://www.zenithmedia.com/smartphone-penetration-reach-66-2018/. Accessed 23 June 2018

Zhang S, McCullagh P, Zhang J, Yu T (2014) A smartphone based real-time daily activity monitoring system. Clust Comput 17(3):711–721

Zhang Y, Xhafa F, Ruiz C, Yao L (2018) Special section editorial: wearable sensor signal processing for smart health. Smart Heal 5–6:1–3 (2018). https://doi.org/10.1016/j.smhl.2018.03.004

Chapter 2
Biomechanical Biofeedback

2.1 Biofeedback

The term *biofeedback* appeared as a scientific research topic in the second half of the 20th century. It was studied first in relation to human physiological processes, where researchers describe its use to treat a variety of conditions and illnesses. For example, Brown (1977) concentrates on stress-related conditions; effects of biofeedback on high blood pressure, epilepsy, and anxiety tension states are studied by Green and Green (1977); Basmajian (1979) discusses biofeedback strategies and applications in treatment of cardiovascular disorders, psychosomatic disorders, psychotherapy, and other disorders.

Shortly afterwards, biofeedback was described in relation to physical medicine and rehabilitation (Fernando and Basmajian 1978), which are based on applying mechanical force, physical exercise, and electrotherapy to remediate mobility impairments and functions. Not much later, the possible benefits of biofeedback started to interest researchers studying physical body motion activity in sports science and biomechanics (Sandweiss 1985; Blumenstein et al. 2002).

The research endeavours in biofeedback related to body motion activity took additional momentum at the beginning of the 21st century with the advancements and wide availability of sensor technology for measuring human motion. A large number of research works have been published in the fields of rehabilitation and sport (Alahakone and Senanayake 2010; Franco et al. 2013; Giggins et al. 2013; Huang et al. 2006; Lee et al. 1996; Paul et al. 2012a, b; Umek et al. 2015).

2.1.1 Definition

Biofeedback is a method of treatment and body control that uses various sensors to measure persons' (*bio*) physiological and physical bodily functions, parameters, and

© Springer Nature Switzerland AG 2018
A. Kos and A. Umek, *Biomechanical Biofeedback Systems and Applications*, Human–Computer Interaction Series, https://doi.org/10.1007/978-3-319-91349-0_2

Fig. 2.1 Biofeedback method. Person's bodily functions, parameters, and activity are sensed and processed. Results are then feed back to the person that tries to act on them

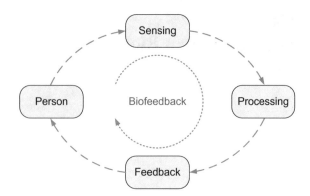

activity of which they are normally unaware or unable to sense by their own senses. Sensors signals and data are processed and the results are communicated back to the person (*feedback*) through one of the human senses (i.e. sight, hearing, touch). The person attempts to act on the received information to change the sensed functions, parameters, and activity in the desired way.

The graphical representation of the above described biofeedback method is shown in Fig. 2.1. It shows a cyclic process where a person is in the loop with elements or functional blocks that are performing sensing, processing, and giving feedback. The biofeedback loop is closed, if the person understands the feedback information and acts accordingly.

2.1.2   Categorization

Biofeedback is categorised into two major categories (Giggins et al. 2013); physiological biofeedback and biomechanical biofeedback, as shown in Fig. 2.2. While physiological biofeedback is connected to physiological processes and states of the human body, biomechanical biofeedback is connected to body activity in the sense of physical movement and states of the human body.

Physiological biofeedback is based on signals and parameters acquired from neuromuscular system, cardiovascular system, respiratory system, brain, skin, and other body systems. Examples of physiological signals are wrist pulse signal, electrocardiograph (ECG) signal, photoplethysmogram (PPG) cardiac cycle signal, electroencephalograph (EEG) signal, functional magnetic resonance images (fMRI), and others. Examples of physiological parameters are temperature, heart rate, blood glucose level, respiratory rate, blood oxygen saturation, and others. It is used primarily in medicine and also in physical rehabilitation.

Biomechanical biofeedback is based on signals and parameters acquired from the measurements of body or body part movements, body posture, and forces produced by the body or forces applied to the body. Examples of biomechanical signals are body

Fig. 2.2 Biofeedback categorization

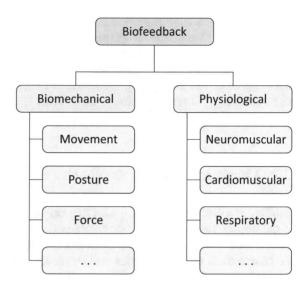

acceleration signal, body rotation speed around its primary axes, body speed, forces produced at body contact with the ground, and others. Examples of biomechanical parameters are body posture at rest, left to right leg force ratio at jumping, centre of mass position, maximal rotation angle around longitudinal body axis in swimming, walking or running symmetry, and many others. It is used primarily in physical rehabilitation and sport.

2.2 Biofeedback Use

Biofeedback is used in a number of fields in medicine, physical rehabilitation, and sport. Biofeedback applications in medicine are aimed at treatment or improvement of medical conditions and illnesses; such as sleep disorders, stress, various addictions, migraine headaches, epilepsy, hypertension, respiratory problems, and many others. Applications in medicine use physiological biofeedback methods and techniques.

Biofeedback applications in physical rehabilitation focus primarily on remediation after illness, injury, and trauma; such as recovery after stroke, muscle dysfunction caused by injury, motor control, and others. Applications in physical rehabilitation employ physiological and biomechanical biofeedback methods and techniques (Alahakone and Senanayake 2010; Franco et al. 2013; Giggins et al. 2013; Giggins et al. 2014; Huang et al. 2006; Kos and Umek 2018; Parvis et al. 2017; Silva 2014).

Biofeedback applications in sport concentrate on motor learning support, performance improvement, and injury prevention; such as human motion analysis, performance measurement and assessment, real-time training systems, tiredness detection, and others. Applications in sport use biomechanical biofeedback methods and techniques (Alahakone and Senanayake 2009; Crowel et al. 2010; Jakus et al. 2017; Kirbi 2009; Kos and Umek 2017; Konttinen et al. 2004; Lieberman and Breazeal 2007; Umek et al. 2015; Umek et al. 2017).

This book focuses on biomechanical biofeedback. In the fields of physical rehabilitation and sport it is tightly connected to motor learning and sports training; therefore we pay special attention to the benefits of biomechanical biofeedback in the above mentioned areas.

2.3   Operation of the Biofeedback Loop

Biofeedback method is essentially a cyclic process, shown in Fig. 2.1, where a person in the loop tries to change one of the body functions, parameters, or activity. Biofeedback is successful, if the change is in the desired way; for example the improvement in the execution of the performed activity. To achieve a beneficial outcome of the biofeedback process, a number of conditions should be met.

2.3.1   Biofeedback Success Conditions

Biofeedback is successful, when the user is appropriately reacting to the given feedback information. Generally that is true, if the feedback loop is closed (the cyclic process from Fig. 2.1 is running) and when the below conditions and requirements are met:

- sensing of the body function, parameter, or activity of interest is possible and available with sufficient accuracy and precision,
- relevant feedback information can be acquired or calculated,
- appropriate feedback type is used,
- feedback information is given in the appropriate modality,
- feedback is available in the proper timing,
- feedback is understandable to the user,
- cognitive load of the person in the loop, as the result of feedback information processing, is not too high.

The above listed conditions and requirements are discussed and explained in the following subsections.

2.3.2   Sensing

Biofeedback process starts with the sensing of the body function, state, parameter, or activity of interest.

Humans have a number of sensors (sense organs) for sensing various quantities in their bodies or in their environment. In addition to the five traditional senses of sight, hearing, touch, smell, and taste, there are other well-known senses, such as senses of balance, pain, temperature, kinesthesia, and vibration. Another group of human senses are physiologically based senses such as, for example, senses of thirst and hunger.

With the development of technology a great number of artificial sensors have been developed and implemented in biofeedback process. Sensors and devices for measuring physiological processes and states of the human body, such as oximeter, EEG, and thermometer, are used in physiological biofeedback. Sensors for measuring physical movement and states of the human body, such as video cameras, inertial sensors, force sensors, and pressure sensors, are used in biomechanical biofeedback.

Humans use the data from their body sensors and from the artificial sensors for perception that allows them to be aware of and to understand themselves and their environment. Perception is an essential element of the biofeedback process.

2.3.3 Feedback Categories

From the perspective of a person in a biofeedback process, feedback can be divided into two main categories: intrinsic feedback and extrinsic feedback. Other feedback categorisations can be done based on the correctness of performed action (positive or negative) and on the returned knowledge information (knowledge of performance or knowledge of result).

Intrinsic Feedback

Intrinsic or inherent feedback is a natural type of feedback that comes from within and uses interoception, proprioception, and exteroception for sensing the body function, state, parameter, or activity.

- *Interoception* is the sense of the internal state of the body (Craig 2003). It is the result of integrating signals from the body allowing for a representation and the determination of the physiological state of the body and the self-awareness.
- *Proprioception* is the sense of the relative position of one's own body or parts of the body and strength of effort being employed in movement. It is the ability to sense stimuli from within the body that help a person to determine position, motion, and equilibrium. For example, even when blindfolded, a person knows through proprioception, if a leg is stretched out in front of the body or behind the body.
- *Exteroception* is the perception of the outside world by sensing the environmental stimuli acting on the body. Common senses used in exteroception are sight, hearing, and touch. For example, a person knows through sight that a ball is approaching his or her head.

Intrinsic feedback is vital in motor control and motor learning (Lauber and Keller 2014). It is also especially important for autonomous skill learners who have reached the level where they know themselves well enough to determine what needs to be corrected purely by the feeling of the skill gathered through interoception, proprioception, and exteroception.

Extrinsic or Augmented Feedback

Extrinsic feedback, also known as augmented feedback, is augmented information provided by an external source, for example by an instructor or some sort of a technical feedback system or device (Magill 2001).

Augmented feedback is information that cannot be acquired without an external source. There are a number of different external sources; from instructors, coaches, teachers, assistants, and trainers to various technical systems that have the capability of sensing, processing and giving augmented feedback. Augmented feedback is received through different modalities, such as vision, audition, tactition, that correspond to senses of sight, hearing and touch, respectively. Augmented feedback is used in both major categories; physiological biofeedback and biomechanical biofeedback. It is particularly popular and useful for accelerated motor learning in physical rehabilitation and sport.

Augmented feedback is used to support intrinsic feedback and is especially important for beginners who have not yet developed the feel of the skill. Augmented feedback decreases the amount of time to master the skill and increases the execution performance level (Schmidt and Wrisberg 2008). Augmented feedback can be positive or negative and it can be presented as knowledge of result or as knowledge of performance (Lauber and Keller 2014).

Other Feedback Categories

Positive feedback is given when a skill is performed correctly. It can be intrinsic or extrinsic and is used to reinforce the correct action. Positive feedback motivates performers to continue correct skill execution or try to raise its performance level. It is also essential for beginners where positive feedback motivates them to continue with the learning process.

Negative feedback is given when a skill is performed incorrectly. It can be intrinsic of extrinsic and it is used to prevent repeating the incorrect action. With autonomous learners the intrinsic negative feedback helps with detecting and correcting their own errors. Beginners can mostly benefit from extrinsic negative feedback, which reduces the probability of developing bad.

Knowledge of performance (KP) is the information that indicates the quality of action. In motor learning KP is concerned about performer's movement pattern. It can be said that KP is based on technique and tells the performer why the movement was correct or incorrect? It can be intrinsic or extrinsic. It is especially important for experienced performers.

Knowledge of result (KR) is extrinsic or augmented information indicating the success of the actions with regard to the set goal. It is given after the performed action. In motor learning KR is related to the outcome of the movement, based on results. Mostly it tells the performer if the movement was successful or unsuccessful. It is extrinsic because it is based purely on the success level of the action. KR can be positive or negative.

2.3.4 Feedback Modalities

Feedback modalities, in regard to biofeedback processes, are human senses that can be used to receive feedback information generated in the biofeedback loop. The most commonly used modalities in biofeedback processes are visual, auditory, and tactile (Sigrist et al. 2013). Feedback is unimodal if it uses only one modality and multimodal if it uses more than one. The correct choice of feedback modality depends primarily on the human sensing ability and occupancy during the task (Jakus et al. 2017). The chosen modality or modalities should not interfere with the performed task and should not cause too much additional cognitive load (Stojmenova et al. 2018). For example, in many cases, visual feedback can interfere with an intensive visual preoccupation with the task and can be rather distracting.

The most commonly used feedback modality is visual. There are countless ways of presenting the visual feedback information; from simple hand-made drawings, body gestures, and semaphores to elaborate and complex displays, performance charts, videos, and 3D animations. The strength and advantage of visual modality is its ability to present large amounts of information in a short time. The drawback of visual modality is its high cognitive load.

Auditory modality uses sound for receiving feedback. Auditory feedback is many times the best choice because this modality is usually unemployed during most tasks (Doza et al. 2011; Eriksson et al. 2011; Schaffert et al. 2010; Vogt et al. 2010). It is also relatively easily implementable and has a low cognitive load.

Tactile feedback modality is becoming an increasingly more appropriate alternative for many applications in sports and physical rehabilitation (Alahakone and Senanayake 2009; Lieberman and Breazeal 2007). This modality has a relatively demanding implementation, and it can be distracting to the user.

2.3.5 Feedback Timing

Feedback timing determines the time when a feedback is given to the person in the biofeedback loop. Two basic timings are defined: concurrent feedback and terminal feedback.

Concurrent feedback is received intrinsically and extrinsically during the task. An example of intrinsic concurrent feedback is sensation of balance during bicycle ride, which helps a person riding the bike make constant subtle moves that prevents them from falling. An example of extrinsic concurrent feedback is the information about the vehicle speed that helps a race driver making an optimal curve.

Terminal feedback is received extrinsically when the task is complete or even some time later. An example of extrinsic or augmented terminal feedback is instructor's advice about possible improvements of the completed task.

A third possibility, not defined until now, is *cyclic feedback*. It is received extrinsically and is defined only for cyclically repeating tasks and it is given immediately after the completion of each individual task cycle. An example of extrinsic cyclic feedback is a technical system that measures the gait parameters, such as step symmetry, and gives feedback about them after each completed step or stride.

Feedback timing is an important parameter of augmented feedback. In the majority of research work about augmented feedback, the feedback information is given with a delay after the performed task. Most of the instructor supported feedback can be classified as terminal feedback. The same is true for the majority of sport applications already available on smartphones. Concurrent feedback has also been found useful for accelerated motor learning in recreational, professional, and amateur sport (Liebermann et al. 2002). While it is not impossible for the coach to give feedback to the athlete during the performed action, such feedback has application only in specific cases. For example, the coach can give concurrent feedback to a boulder climber during the action, but concurrent feedback would be of no value to the athlete performing a high jump.

2.4 Example—Motor Learning with Augmented Feedback

This section is intended to present the biofeedback method through an example of motor learning in sports supported by biomechanical biofeedback with augmented feedback. To better understand the purpose of the example, motor learning basics are described first, followed by three exemplar biomechanical biofeedback systems, each supporting a specific learning scenario.

2.4.1 Motor Learning Basics

Motor learning is a large interdisciplinary research area that encompasses a number of scientific fields, such as psychology, medicine, physical rehabilitation, sport, and others. The purpose of this section is to briefly discuss the fundaments of motor learning that are the basis for a number of topics in this and later chapters of this book.

Motor learning is a change, resulting from practice. It is a relatively permanent skill as the capability to respond appropriately is acquired and retained.

Motor learning is a process of learning new movements and it is based on repetition and practice (Bilodeau et al. 1969). Numerous correct executions, mostly several thousand of them, are required to adequately learn a certain movement. According to sports experts, feedback is the most important concept for motor learning, except the practice itself (Bilodeau et al. 1969).

Law of Effect and Knowledge of Result

Motor learning has its origins in the *law of effect*, which was proposed as early as 1898 by Thorndike. Law of effect states that, the association between some action and some environmental condition is enhanced when the action is followed by a satisfying outcome. On the other hand, a dissatisfying outcome weakens the association.

- *Enhancing the association*—if a child motions its hands and legs in just the right way, it performs a crawling motion. Since the child increases its mobility, this is considered as a satisfying outcome. Because of the satisfying outcome, the association between this particular arm and leg motions and being on all fours is enhanced.
- *Weakening the association*—if a child contracts certain muscles, resulting in a painful fall, the child will decrease the association between these muscle contractions and the environmental condition of standing on two feet.

The law of effect cannot explain all the aspects of motor learning. The author of (Adams 1971) demonstrates that knowledge of result, especially its guidance role, is equally influential in motor learning.

Stages of Motor Learning

Motor learning is a process that takes time and effort. It may take several thousand repetitions and several months to master a particular skill. Motor learning process is divided into three phases (Adams 1971; Motor skill 2018):

- *Cognitive phase*—the learner is new to a specific task and the learning process starts with a primary thought "What needs to be done?"

  - Considerable cognitive activity is required for the determination of appropriate strategies for reaching the desired outcome.
  - Strategies yielding favourable outcomes are retained and inappropriate strategies are discarded.
  - The performance can be greatly improved in a short amount of time.

- *Associative phase*—the learner has established the most effective way of task performance and starts to introduce subtle adjustments in performance.

  - Movements become more consistent and improvements are more gradual.
  - This phase can last for quite a long time.
  - The skills in this phase are fluent, efficient and aesthetically pleasing.

- *Autonomous phase*—this phase may take very long time to reach; from several months to several years.

  - The performer *automatically* completes the task without having to pay any attention to performing it.

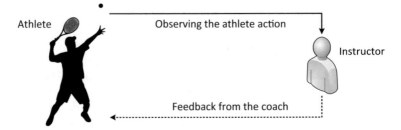

Fig. 2.3 Instructor based augmented feedback motor learning without the use of technical equipment. The instructor is monitoring athlete's actions and gives the feedback about the performance, results, and advice about possible improvements

2.4.2 *Biomechanical Biofeedback with Augmented Feedback*

Biomechanical biofeedback can be beneficially used in motor learning. According to sports experts, feedback is the most important concept for motor learning, except the practice itself (Bilodeau et al. 1969). It can be concluded that motor learning heavily depends on the feedback given to the learners (Adams 1971; Bilodeau et al. 1969; Eriksson et al. 2011; Liebermann et al. 2002; Motor skill 2018; Paul et al. 2012a, b; Sigrist et al. 2013). Here we present the accelerated motor learning possibilities with the aid of biomechanical biofeedback through the examples in sport.

During the practice, inherent (natural) feedback information is provided internally through human sense organs. In addition to natural feedback channels, augmented feedback can be provided by an external source; traditionally by instructors and trainers, and recently also by technical equipment, devices, and advanced feedback systems.

Motor learning includes various forms of natural feedback. Most of these forms are acquired during early childhood motor development. In the case of natural feedback, one is limited to the capabilities of human senses and cognition. Motor learning, especially in sports, can be accelerated with the aid of an instructor. The instructor follows or monitors athlete's actions and gives the feedback about the performance, results, and also advice about possible improvements—the instructor is providing the augmented feedback. With this type of augmented feedback technical equipment is not necessary because the instructor is performing all the tasks of the feedback system (sensing, processing, giving feedback). Instructor based feedback without the use of technical equipment is presented in Fig. 2.3. Instructor based augmented feedback system is particularly useful in the cognitive and also in the associative motor learning phase, as described in Sect. 2.4.1.

The traditional way of instructor based motor learning can be improved by introducing technical equipment that is capable of measuring, calculating and presenting the properties of the performed action. In Fig. 2.4 the technical equipment is represented by sensing, processing, and monitoring elements. The main reason for using technical equipment is the possibility to obtain information that is out of reach or

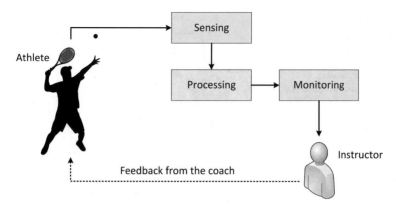

Fig. 2.4 Instructor based augmented feedback motor learning with the use of technical equipment. A biofeedback system is monitoring athlete's actions. The instructor gives the feedback to the athlete based on his/her own observation or based on the information from the technical equipment of the biofeedback system

beyond human sensing capabilities. For example, an instructor cannot "see" the level of force an athlete is exerting during the jumps or see the exact spot where a tennis ball hits the racket during a serve.

Specialized technical equipment can measure, calculate, and present the athlete force and the tennis ball hitting spot. For example, in Fig. 2.4 the sensing can be done by a high speed, high definition camera that is recording the tennis serve. A streamed video is processed and the ball hitting spot is calculated. The instructor gets a graphical representation of the serve, accompanied by several other relevant parameters, on a tablet screen. The coach can then analyse the data and give advice to the tennis player.

In addition to natural and instructor based feedback channels, artificial feedback channels are also possible. Artificial channel augmented feedback exploits the human sensory systems; vision, hearing, and touch.

A technology based motor learning solution with artificial channel feedback is shown in Fig. 2.5. This solution does not include an instructor. In such biofeedback systems athletes have attached sensors that measure body actions. Sensor signals are processed and the results are communicated back to the person (feedback) through one of the human senses (i.e., sight, hearing, touch) (Sigrist et al. 2013). Athletes attempt to act on received information to change the body motion as desired. Such motor learning methods are suitable for recreational, professional, and amateur users (Liebermann et al. 2002; Sigrist et al. 2013). Autonomous augmented feedback system is particularly useful in the autonomous motor learning phase, as described in Sect. 2.4.1

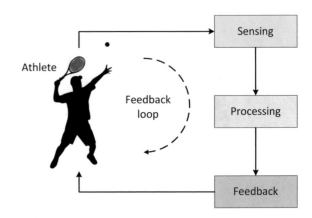

Fig. 2.5 Autonomous augmented feedback during the technology supported motor learning. A biofeedback system is monitoring athlete's actions and gives the feedback about the performance directly to the athlete. The feedback loop is closed when the user tries to adjust the performed action according to the given feedback

2.5 Benefits and the Need for Augmented Biofeedback

While there are no doubts about the benefits of natural biofeedback, some questions are still remaining; what are the cardinal advantages and benefits of the augmented biofeedback; is augmented feedback always beneficial? As it has already been shown, feedback motivates, reinforces and speeds learning process, but is this always true? Some general guidelines are collected in the below list.

- Augmented feedback is essential when:

  - Intrinsic feedback information is not available (cannot see the entire environment needed for performing the task).
  - Person's modalities are impaired due to injury, age, or illness.
  - Intrinsic feedback is available, but the performer is inexperienced and cannot interpret it.

- Augmented feedback is not needed when:

  - Feedback is inherently provided by the task itself; for example, the performer can see the result of shooting on the target.
  - In observational learning in situations without the need for augmentation.

- Augmented feedback can enhances skill acquisition in any relatively complex skill that requires many repeated attempts to achieve a certain degree of success in achieving performance goals.
- Augmented feedback obstruct skill learning when:

  - Beginners acquire dependency on given augmented feedback that will later not be available in real situations.
  - The given augmented feedback is improper or erroneous.

What about the augmented feedback backed by the technology? As it will be shown in the following chapters of this book, technology offers the possibility to obtain information that is out of reach of human senses or the information that is beyond human senses capabilities. It is not arguable that technology can outperform human senses (from both, the performer and the instructor) in the variety of quantities measured, precision, speed, detail of measurement, etc. It is also safe to state that technology gives objective results, while humans make subjective assessments. This book is trying to answer the above questions and doubts through presentation of biofeedback systems and applications based on the state-of-the-art technology.

Note

This chapter offers the background of biofeedbackBiofeedback and of some related topics, such as motor learningMotor learning, that are essential for the understanding of the remainder of the book. The content of the following chapters narrows the scope to the biomechanical biofeedbackBiomechanical biofeedback category of biofeedbackBiofeedback and to the technology based augmented feedbackFeedback. Emphasis is also on systems and applications that are able to provide concurrent feedbackConcurrent feedback.

> In the remainder of this book the word *biofeedback* denotes *biomechanical biofeedback*.

References

Adams JA (1971) A closed-loop theory of motor learning. J Motor Behav 3(2):111–150

Alahakone AU, Senanayake SA (2009, December) A real time vibrotactile biofeedback system for improving lower extremity kinematic motion during sports training. In: International conference of soft computing and pattern recognition, 2009. SOCPAR'09. IEEE, pp 610–615

Alahakone AU, Senanayake SA (2010) A real-time system with assistive feedback for postural control in rehabilitation. IEEE/ASME Trans Mechatron 15(2):226–233

Basmajian JV (1979) Biofeedback: principles and practice for clinicians. Williams & Wilkins

Bilodeau EA, Bilodeau IM, Alluisi EA (1969) Principles of skill acquisition. Academic Press

Blumenstein B, Bar-Eli M, Tenenbaum G (2002) Brain and body in sport and exercise—biofeedback applications in performance enhancement. Wiley

Brown BB (1977) Stress and the art of biofeedback. Harper & Row

Craig AD (2003) Interoception: the sense of the physiological condition of the body. Curr Opin Neurobiol 13(4):500–505

Crowell HP, Milner CE, Hamill J, Davis IS (2010) Reducing impact loading during running with the use of real-time visual feedback. J Orthop Sports Phys Ther 40(4):206–213

Dozza M, Chiari L, Peterka RJ, Wall C, Horak FB (2011) What is the most effective type of audio-biofeedback for postural motor learning? Gait Posture 34(3):313–319

Eriksson M, Halvorsen KA, Gullstrand L (2011) Immediate effect of visual and auditory feedback to control the running mechanics of well-trained athletes. J Sports Sci 29(3):253–262

Fernando CK, Basmajian JV (1978) Biofeedback in physical medicine and rehabilitation. Biofeedback Self-Regul 3(4):435–455

Franco C, Fleury A, Guméry PY, Diot B, Demongeot J, Vuillerme N (2013) iBalance-ABF: a smartphone-based audio-biofeedback balance system. IEEE Trans Biomed Eng 60(1):211–215

Giggins OM, Persson UM, Caulfield B (2013) Biofeedback in rehabilitation. J Neuroeng Rehabil 10(1):60

Giggins OM, Sweeney KT, Caulfield B (2014) Rehabilitation exercise assessment using inertial sensors: a cross-sectional analytical study. J Neuroeng Rehabil 11(1):158

Green E, Green A (1977) Beyond biofeedback. Delacorte

Huang H, Wolf SL, He J (2006) Recent developments in biofeedback for neuromotor rehabilitation. J Neuroeng Rehabil 3(1):11

Jakus G, Stojmenova K, Tomažič S, Sodnik J (2017) A system for efficient motor learning using multimodal augmented feedback. Multimed Tools Appl 76(20):20409–20421

Kirby R (2009) Development of a real-time performance measurement and feedback system for alpine skiers. Sports Technol 2(1–2):43–52

Konttinen N, Mononen K, Viitasalo J, Mets T (2004) The effects of augmented auditory feedback on psychomotor skill learning in precision shooting. J Sport Exerc Psychol 26(2):306–316

Kos A, Umek A (2017) Smart sport equipment: SmartSki prototype for biofeedback applications in skiing. Pers Ubiquitous Comput, 1–10

Kos A, Umek A (2018) Wearable sensor devices for prevention and rehabilitation in healthcare: swimming exercise with real-time therapist feedback. IEEE Internet Things J. https://doi.org/10.1109/jiot.2018.2850664

Lauber B, Keller M (2014) Improving motor performance: selected aspects of aug-mented feedback in exercise and health. Eur J Sport Sci 14(1):36–43

Lee MY, Wong MK, Tang FT (1996) Using biofeedback for standing-steadiness, weight-bearing training. IEEE Eng Med Biol Mag 15(6):112–116

Lieberman J, Breazeal C (2007, April) Development of a wearable vibrotactile feedback suit for accelerated human motor learning. In: 2007 IEEE international conference on robotics and automation. IEEE, pp 4001–4006

Liebermann DG, Katz L, Hughes MD, Bartlett RM, McClements J, Franks IM (2002) Advances in the application of information technology to sport performance. J Sports Sci 20(10):755–769

Magill R (2001) The effect of augmented feedback on skill learning. Motor Learn Concepts Appl, 235–245

Motor skill (2018) https://en.wikipedia.org/wiki/Motor_skill. Accessed 2 Jun 2018

Parvis M, Corbellini S, Lombardo L, Iannnucci L, Grassini S, Angelini E (2017, May) Inertial measurement system for swimming rehabilitation. In: 2017 IEEE international symposium on medical measurements and applications (MeMeA). IEEE, pp 361–366

Paul M, Garg K, Sandhu JS (2012a) Role of biofeedback in optimizing psychomotor performance in sports. Asian J Sports Med 3(1):29

Paul M, Garg K, Sandhu JS (2012b) Role of biofeedback in optimizing psychomotor performance in sports. Asian J Sports Med 3(1):29

Sandweiss JH (1985) Biofeedback and sports science. In: Biofeedback and sports science. Springer, Boston, MA, pp 1–31

Schaffert N, Mattes K, Effenberg AO (2010) A sound design for acoustic feedback in elite sports. Auditory display. Springer, Berlin, Heidelberg, pp 143–165

Schmidt RA, Wrisberg CA (2008) Motor learning and performance: a situation-based learning approach. Human Kinetics

Sigrist R, Rauter G, Riener R, Wolf P (2013) Augmented visual, auditory, haptic, and multimodal feedback in motor learning: a review. Psychon Bull Rev 20(1):21–53

Silva ASM (2014) Wearable sensors systems for human motion analysis: sports and rehabilitation. Doctoral dissertation, Universidade do Porto (Portugal)

Stojmenova K, Duh SE, Sodnik J (2018) A review on methods for assessing driver's cognitive load. IPSI BGD Trans Internet Res 14(1)

Umek A, Kos A, Tomazic S (2017) Biomehanska povratna zanka z delovanjem v realnem času. Elektrotehniski Vestnik 84(1/2):1

Umek A, Tomažič S, Kos A (2015) Wearable training system with real-time biofeedback and gesture user interface. Pers Ubiquit Comput 19(7):989–998

Vogt K, Pirrò D, Kobenz I, Höldrich R, Eckel G (2010) PhysioSonic-evaluated movement sonifi-cation as auditory feedback in physiotherapy. Auditory display. Springer, Berlin, Heidelberg, pp 103–120

Chapter 3
Biofeedback System

3.1 Background

Biofeedback basics and related topics, essential for the understanding of the following chapters of the book, are given in Chap. 2. From this point on, the book narrows its scope to the presentation and discussion of the most important aspects of the biomechanical biofeedback systems, which are implemented with the support of technical equipment and systems. In the great majority of cases such systems use technology to implement functions of sensing, processing, and feedback, as defined in Fig. 2.1. In some cases, some of the above mentioned functions may be performed by an instructor.

In other words, the book studies biomechanical biofeedback systems with augmented feedback that is provided by technical equipment. In the remainder of this book the word *biofeedback* denotes biomechanical biofeedback systems that give augmented feedback; for clearer expression and better readability.

The primary focus of the book is kept on the technological perspective of biofeedback systems; user's (person's) perspective comes secondary, but is by no means unimportant. Some more emphasis is given to the subgroup of real-time biofeedback systems that provide concurrent feedback (Alahakone and Senanayake 2010; Umek et al. 2015).

As it can be learnt from Fig. 2.1, the biofeedback method diagram is relatively simple. But, when given a number of different options for the elements of sensing, processing, and feedback, implementations can vary greatly depending on the combination used. For example, sensing and processing is done by the technical equipment, but feedback is given by the instructor, as shown in Fig. 2.4. This book presents and studies a number of different biofeedback systems where any of the functional blocks from Fig. 2.1 (sensing, processing, and feedback) can be implemented by technical equipment or by an instructor. Much more stress is given to the discussion of the technical equipment based systems than to the instructor based systems because the later have been well studied in sports and similar research fields.

© Springer Nature Switzerland AG 2018
A. Kos and A. Umek, *Biomechanical Biofeedback Systems and Applications*,
Human–Computer Interaction Series, https://doi.org/10.1007/978-3-319-91349-0_3

Table 3.1 Implementations of biofeedback method functional blocks from Fig. 2.1

Function executant	Functional block		
	Sensing	Processing	Feedback
Instructor	Sight	Consideration	Oral advice
	Hearing	Experience	Body gestures
	Touch	Consultation	Graphical
	. . .	. . .	Numerical
			. . .
Technical equipment	Motion sensors	Mechanical	Screen
	Photography	Analogue electronics	Loudspeaker
	Video camera	Microcontroller	Headphones
	Force plate	Smartphone	Vibration device
	Pressure sensor	Computer	. . .
	. . .	. . .	

Table 3.1 summarizes the most common implementations of functional blocks of the biofeedback method from Fig. 2.1. The table lists implementations of the combination of the functional block (sensing, processing, and feedback) and function executant (instructor, technical equipment). While the instructor performs the sensing through his/her senses, technical equipment employs a great variety of different sensors. Technical equipment based processing can be done by a large number of devices; from mechanical devices calculating force from the measured pressure, to supercomputers performing deep learning on large amounts of measured data. Instructors do the processing based on consideration, experience, consultation, comparison, and other techniques. They give feedback orally, by gestures, graphically by drawing, etc. Technical equipment used to give feedback includes various screens, speakers, vibration devices, and other devices that are able to present the feedback information in any of the human modalities suitable for biofeedback.

Almost any combination of functional blocks and their executants from Table 3.1 is possible. For example, an instructor may use a force plate to measure forces during the movement, he/she processes the gained signals and/or measured parameters based on his/her experience, finally, the feedback is given by pressing the button that causes a beep notifying the performer about the incorrect movement. A complemental case is when an instructor estimates the parameters of the movement and enters them into the computer for processing and analysis. The instructor then orally gives feedback about the results.

3.2   Architecture

A general architecture of the biomechanical biofeedback system is presented in Fig. 3.1 (Baca et al. 2009; Biofeedback 2018; Huang et al. 2006; Kos et al. 2018a, b; Umek et al. 2015; Umek and Kos 2016a, b). When comparing it with Fig. 2.1, it can be noticed that sensing, processing, and feedback functional blocks from Fig. 2.1 are changed to sensor, processing, and feedback *devices* in Fig. 3.1. Also, a *person* from Fig. 2.1 is denoted as a *user* in Fig. 3.1. This denomination is kept throughout the book in general representations of biofeedback systems because the person is treated as the user of the *system*. In some special-purpose biofeedback systems the *person* is denoted as an *athlete* (sport) or as a *patient* (rehabilitation). Similarly, an *instructor* (Fig. 2.4) can be denoted as a *coach*, as a *trainer* or as a *therapist*.

Biofeedback system depicted in Fig. 3.1 implements a biofeedback loop with sensors, processing device and feedback device that provide sensing, processing, and feedback functionalities. Between these devices and the user are communication channels that enable information flow. The user communicates with sensors by performing an action that is sensed and measured; for example, by moving a leg.

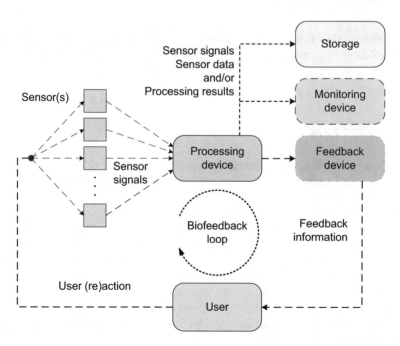

Fig. 3.1 Biomechanical biofeedback system with augmented feedback provided by technical equipment. Sensors feed their signals to the processing device for analysis. Feedback device gives the calculated feedback information to the user through one of human modalities. User (re)action alters sensor signals, thus closing the feedback loop of the system. Monitoring device and storage are optional

Sensors send the acquired signals and data about the leg movement to the processing device though the communication channel implemented in one of the available technologies; for example, through Bluetooth wireless technology. Processing device performs calculations and forwards results to the feedback device; for example, information about the error in the performed leg movement. Feedback device conveys the feedback information to the user throughout one of the human modalities; for example, it makes a sound that notifies the user about the error found in the performed leg movement. The user reacts to the given feedback; for example, after hearing the beep and interpreting its meaning, the user reacts by changing the leg movement pattern.

In addition of the above described elements, that are essential for the biofeedback loop operation, monitoring device and data storage can also be included into the biofeedback system. They are not a part of the biofeedback loop, but they are indispensable especially in systems with terminal feedback, where a user or an instructor can thoroughly analyse the recorded signals and can therefore most likely acquire better feedback information.

3.3   System Elements

Biofeedback systems conformable to Fig. 3.1 include sensors, processing devices, feedback devices or actuators, and users that are interconnected by communication channels. Monitoring device and storage are optional.

3.3.1   Sensor(s)

Sensors detect actions, state, and parameters of users as well as various parameters of their physical environment. Sensors produce signals that are sent to the processing device for analysis.

Sensors in biofeedback systems are mostly attached to the users' body or integrated into the equipment used, but they can also be in the vicinity of users or even implanted into their bodies. Accelerometers and gyroscopes in the wrist band are an example of user-attached sensors (Umek and Kos 2016a, b, 2018), flex sensors integrated into the smart ski are an example of equipment-integrated sensors (Kos and Umek 2017), high speed cameras and radars installed on the stadium are an example of the sensors in the vicinity of users, and glucose sensors are an example of the implanted sensors. Each biofeedback system may include one or more sensors of the same or different types.

There are a large number of different sensors measuring a number of different quantities. The most important properties of sensors are: dynamic range, resolution, sampling frequency, precision, accuracy, and sensitivity. The study of sensors and their properties is a very important topic within the biofeedback system research.

3.3.2 Processing Device

The processing device is the core of the biofeedback system. It receives sensor signals and data, analyses them, and generates feedback signals that are sent to a feedback device.

The processing device can be a separate device (i.e., laptop or tablet), a component of a multifunction body-attached device (i.e., smartphone or smart watch), a dedicated wearable processing device (i.e. micro-controller), or a virtual device (i.e., cloud virtual appliance). The feedback information can be expressed in the same quantity as measured variable; for example, showing the measured acceleration signal during running. This is called a *direct feedback*. The other option is expressing the feedback information in some other quantity; for example, the measured acceleration signal is processed and the user is given the running frequency or step duration symmetry as the feedback information. This is called a *transformed feedback* (Giggins et al. 2013). While the direct feedback does not need a lot of processing, the transformed feedback processing efforts can range from low to very high. For example, not a lot of processing is needed to calculate step duration, but analysing the difference in the acceleration pattern during the course of a running exercise and simultaneously comparing it previous exercises, requires a lot of processing power from the processing device (Umek and Kos 2016a, b).

3.3.3 Feedback Device (Actuator)

Feedback device uses human senses to communicate feedback information to the user. Feedback devices are also called actuators. Actuators are devices that actuate or incite action of a user with their activity.

The most commonly used senses are hearing, sight, and touch; commonly referred to as auditory, visual, and haptic modalities. For example, headphones are an actuator that uses auditory modality for feedback.

Similarly to sensors, each biofeedback system may include one or more feedback devices of the same or different type and they can be placed in the vicinity of the user or attached to the user's body. A large display showing user's results, an actuator using the visual modality, is usually placed near the user. A vibrotactile actuator, using the haptic modality, must be attached to the appropriate body part of the user.

3.3.4 Users and Communication Channels

Users and communication channels are important biofeedback systems elements. User's (re)actions are necessary to close the feedback loop. Communication channels are needed first for the transmission of sensor signals to the processing device, secondly for the transfer of feedback information to the feedback device, and thirdly for conveying the feedback device stimuli to the human senses. The first two communication channels are implemented based on wired or wireless communication technologies; the last one is a natural channel belonging to one of the human modalities. Although wired technologies can be used in practice, e.g., to send a feedback signal from the body-attached processing device to the nearby vibrotactile actuator (Lieberman and Breazeal 2007), wireless communication technologies are most commonly used (Baca et al. 2009; Kos et al. 2018a, b; Umek and Kos 2016a, b). A more detailed study of communication technologies is given in Sect. 6.6.3 of this book.

3.3.5 Optional Elements

All of the above described elements of the biofeedback system are necessary for its correct operation. Apart from those elements, Fig. 3.1 includes also a monitoring device and storage. Both of them are optional. The most common monitoring devices are screens in all forms, from smart watches and smartphones to computer monitors and big screen TVs.

The *monitoring device* is designed to monitor sensor signals and analysis results. It is used by the instructor and/or the user. The instructor can monitor sensor signals and processing results concurrently with the performed task and it can be a source of information for giving feedback to the user. In the vast majority of cases, users monitor sensor signals and processing results after the completed task as during the task they are preoccupied by the task itself.

Storage is a place where biofeedback systems can store their data. It can be the local memory of the processing device, local disk, external disk, or even cloud storage (Badawi et al. 2017; Kos et al. 2018a, b; Umek et al. 2015). Data stored includes, but is not limited to, sensor signals, processing results, feedback information, biofeedback system settings and parameters, user data, etc. Data can be stored in the form of binary or text files, tables or databases, or any other form defined by the biofeedback system. Stored data can be particularly useful for long term historic analysis of collected data, such as statistics and trends of the individuals, groups, and populations. If the amount of collected data is large enough, data mining and machine learning techniques can also be used (Kos et al. 2018a, b).

3.4 System Operation

The general operation of the biofeedback system is presented in Sect. 2.1, with the study of its success conditions, sensing possibilities, feedback types, feedback modalities, and feedback timing. The specifics of the biomechanical biofeedback systems with augmented feedback provided by technical equipment, shown in Fig. 3.1 and presented in Sect. 3.2 and 3.3, are discussed next.

The operation of the biofeedback system is based upon the biofeedback loop illustrated in Fig. 3.1. Sensors continuously send signals to the processing device for analysis. The feedback device is driven the result of the analysis of the sensor signals and application algorithms. The activity of the feedback device, and thus the activity of the feedback loop, is defined by these feedback signals. The user tries to act on the signals from the feedback device, thus closing the biofeedback loop.

3.4.1 Operation Modes

Depending on the system state and the actions of the user, the biofeedback loop operates in the following modes.

- *Standby*: The processing device continuously processes sensor signals, but no feedback signal is generated; the feedback device is not active, and thus, the biofeedback loop remains open.
- *User guidance*: Feedback signal is constantly generated by the processing device to guide the user in performing a certain movement or assuming a certain position; in this mode, the loop remains constantly closed, and the user is expected to react concurrently.
- *Error detection*: Feedback signal is generated only when the system detects an unwanted action or state of the user; in this mode, the loop is closed for shorter periods of time, and during those periods the user attempts to react concurrently.

The (re)action of the user depends on the perceived feedback information and the operation mode of the biofeedback loop. In the *standby* mode, the user is not expected to react, and thus, the feedback device is not active; not giving feedback information to the user. In the *user guidance* mode, the user is expected to constantly follow the communicated biofeedback information and react concurrently. In the *error detection* mode, the user can either immediately react by attempting to correct or abort the action in progress or, more likely, the user will attempt to avoid errors in subsequent repetitions. In certain cases, an immediate reaction is not desirable or might even be impossible or harmful.

3.4.2 Timing

The definition of timing in the biofeedback system is given in Sect. 2.3.5, where three types of feedback are defined: two basic and well known (a) concurrent feedback, (b) terminal feedback, and a newly defined (c) cyclic feedback.

In biomechanical biofeedback systems with augmented feedback provided by technical equipment the possible feedback timing depends also on the processing capabilities of the biofeedback system (Umek et al. 2016a, b).

- When the feedback is given concurrently, the entire processing must be performed in real time. Such systems are denoted as *real-time biofeedback systems.*
- When the feedback is terminal, given after the action is completed, then the system can afford to do the processing after the action. Such systems are denoted as *post-processing biofeedback systems.*
- When the feedback is given after each completed cycle of the action, the system must do the processing of the entire block of data collected during the time frame of the cycle until the end of each cycle or immediately after it. Such systems are denoted as *block-processing biofeedback systems.*

While the post-processing operation of the system does not represent a computational problem to the most of the processing devices, real-time processing and block-processing operation of the system can be a challenging problem.

For example, in the post processing operation of the system all the processing must be finished at a specified time after the completion of the activity. This time delay can vary from a few seconds, to minutes, hours, days or even longer time periods, depending on the intended ways of use. In the real-time operation of the system the processing device has to finish the processing within the time frame of one sensor sampling period, which can be as low as one millisecond or even less. Similarly, in the block-processing operation of the system the processing device has to finish the processing within the time frame of one cycle, which varies from action to action. The cycle time frame is typically in the range between one and a few seconds, but can also be less or more.

While real-time processing and block processing are generally not a problem for devices with relatively high computation power such as laptops, desktops, and cloud based computing, it can prove too demanding for mobile and/or embedded devices such as microcontrollers and smartphones.

References

Alahakone AU, Senanayake SA (2010) A real-time system with assistive feedback for postural control in rehabilitation. IEEE/ASME Trans Mechatron 15(2):226–233

Baca A, Dabnichki P, Heller M, Kornfeind P (2009) Ubiquitous computing in sports: a review and analysis. J Sports Sci 27(12):1335–1346

Badawi HF, Dong H, El Saddik A (2017) Mobile cloud-based physical activity advisory system using biofeedback sensors. Future Gener Comput Syst 66:59–70

Biofeedback (2018) https://en.wikipedia.org/wiki/Biofeedback. Accessed 4 Jun 2018

Giggins OM, Persson UM, Caulfield B (2013) Biofeedback in rehabilitation. J Neuroeng Rehabil 10(1):60

Huang H, Wolf SL, He J (2006) Recent developments in biofeedback for neuromotor rehabilitation. J Neuroeng Rehabil 3(1):11

Kos A, Umek A (2017) Smart sport equipment: SmartSki prototype for biofeedback applications in skiing. Pers Ubiquitous Comput, 1–10

Kos A, Milutinović V, Umek A (2018a) Challenges in wireless communication for connected sensors and wearable devices used in sport biofeedback applications. Future Gener Comput Syst

Kos A, Wei Y, Tomažič S, Umek A (2018b) The role of science and technology in sport. Procedia Comput Sci 129:489–495

Lieberman J, Breazeal C (2007, April) Development of a wearable vibrotactile feedback suit for accelerated human motor learning. In: 2007 IEEE international conference on robotics and automation. IEEE, pp 4001–4006

Umek A, Kos A (2016a) The role of high performance computing and communication for real-time biofeedback in sport. Mathemat Probl Eng 2016

Umek A, Kos A (2016b) Validation of smartphone gyroscopes for mobile biofeedback applications. Pers Ubiquit Comput 20(5):657–666

Umek A, Kos A (2018) Wearable sensors and smart equipment for feedback in watersports. Procedia Comput Sci 129:496–502

Umek A, Tomažič S, Kos A (2015) Wearable training system with real-time biofeedback and gesture user interface. Pers Ubiquit Comput 19(7):989–998

Chapter 4
Biofeedback System Architectures

4.1 Implementation Diversity

The diversity of possible applications of biofeedback systems is very high. Therefore flexible system architecture is proposed that supports implementations based on a variety of system constraints and demands. The final implementation of the system depends on many factors; among them the most influential tend to be the intended application scenario and the physical extent of the space in which the system is used.

In all versions of the system, the biofeedback loop must be closed; hence the system devices that participate in the loop must communicate over some communication channels. For example, in a real-time biofeedback system the employed communication channels must have low latency.

In the majority of biofeedback system implementations, sensors are attached to the user, and the feedback device (actuator) is located close to the user or attached to the user. The location of sensors and actuators is the primary condition affecting the choice of the processing device location; it can be on the user, close to the user, or at any other location away from the user.

In the following sections of this chapter, several architectures of the biofeedback system are presented. They are defined based on the constraints and the demands of various biofeedback system implementations.

4.2 Constraints

Biofeedback systems are subjected to a number of constraints that define and bound their operation. The most important constraints in biomechanical biofeedback systems with augmented feedback are related to space, time, and computational power. Other constraints, such as energy and sensor accuracy, are less aggravating because efficient mechanisms for their mitigation exist.

© Springer Nature Switzerland AG 2018   49
A. Kos and A. Umek, *Biomechanical Biofeedback Systems and Applications*,
Human–Computer Interaction Series, https://doi.org/10.1007/978-3-319-91349-0_4

4.2.1 Space Constraint

The space constraint defines the possible locations of biofeedback system elements (Umek et al. 2015). We define three basic types of biofeedback systems: (a) *personal space system*, where all system elements are attached to the user, (b) *confined space system*, where the elements are distributed within a defined and limited space, and (c) *open space system*, where elements are not restricted in space.

In most cases sensors and actuators are attached to the user. The most diverse is the location of the processing device that can be attached to the user, close to the user or at any other location connected to the system.

4.2.2 Time Constraint

A biofeedback system can work only, if the feedback loop is closed. That means that the user receives, understands, and reacts to the given feedback information. The feedback information can be given at different times: (a) *terminal feedback* is given after the activity has been performed; (b) *concurrent feedback* is given during the activity; (c) *cyclic feedback* is given after each completed cycle of the cyclic activity.

An important parameter is the feedback loop delay that consists of communication delays for the transmission of sensor and feedback signals, processing delay, and user reaction delay. Acceptable feedback loop delay depends on the basic type of feedback. For the terminal feedback the timing is not that important; it can vary from a few seconds (instructor advice given immediately after the activity has been completed) to a few hours or even days (video analysis of a recorded activity). For the concurrent feedback, the delay in the technical part of the feedback loop (communication and processing delay) should be shorter than the human reaction time (Kos et al. 2018). For the cyclic feedback, the delay in the technical part of the feedback loop can be on principle longer than the delay for the concurrent feedback, but it should be low enough to assure the prompt sensor signal transmission and the timely feedback information transfer to the user after its computation in each activity cycle is completed.

4.2.3 Computation Constraint

Computation constraint is closely related and dependent on the space and time constraints as well as on the properties of sensors and actuators. Processing in the biofeedback loop can be done in: (a) *post processing* mode, where sensor signals acquired during the performed activity are processed after the activity finishes; (b) *real-time* mode, where sensor signals are processed concurrently with the execution of the activity; and (c) *block processing* mode, where the system must do the processing of the entire block of data collected during the time frame of the cycle until the end of each cycle or immediately after it.

While the post processing mode does not represent a computational problem to the most of the processing devices and communication technologies, real time operation mode can many times be a very difficult problem. Block operation mode is somewhere in-between of both abovementioned modes.

For example, in the post processing mode the absolute communication delay is not of paramount importance providing that the processing device receives and analyses all the data at a specified time after the completion of the activity. This time delay can vary from a few seconds, to minutes, hours, days or even longer time periods, depending on the intended ways of use. In the real time mode the processing device has to finish the processing within the time frame of one sensor sampling period, which can be as low as 1 ms or even less. Similarly, in the block processing mode some operations, such as triggering event processing, may be required to finish within the time frame of one sensor sampling period. Other operations and algorithms, such as the average speed during the cycle, may be required to finish shortly after the cycle completion.

While real time and block processing is generally not a problem for devices with relatively high computation power such as laptops, desktops, and cloud based computing, it can prove too demanding for mobile and/or embedded devices such as microcontrollers and smartphones.

4.2.4 Other Constraints

Apart from the abovementioned constraints, biofeedback systems also have a number of other constraints. Among them the most important are energy and accuracy constraints, which are characteristic especially in wearable biofeedback systems.

Energy constraint limits the autonomy of sensors, actuators, and processing devices. This is especially important in (a) wireless sensor based networks, where sensors can be in an inaccessible or very remote location or (b) in medical use, where sensors can be implanted into the human body. In wearable biofeedback systems this constraint has only very limited influence as sensor or sensor, actuators, or processing device battery is easily accessible, therefore it can also be easily recharged or replaced when needed.

Accuracy constraint limits the usability and validity of measured quantities acquired by sensors and results derived from them. While sensor inaccuracies may be critical for some applications, that is rarely true for biofeedback applications. The main reasons are: (a) in most activities frequent and high quality sensor calibration can be performed periodically or on as-needed basis (Kos et al. 2016), (b) accurate activity measurements in biofeedback systems are mostly performed in short time frames, where sensor inaccuracies do not exceed the required accuracy threshold (Umek and Kos 2016).

4.3 Properties

Numerous biofeedback system architectures can be devised based on different combinations of space, time and computation constraints. While these constraints define the base architecture of the system, its details are defined by its properties, such as structure, functionality, and physical extent.

4.3.1 Structure

In the terms of *structure* biofeedback systems can be divided into compact and distributed system types.

- In *compact biofeedback systems* all of its elements (sensors, actuators, processing devices, and communication channels) are very close to each other; integrated into one device or attached to the same person. For example, a smartphone can be used as a compact biofeedback system. It integrates sensors, processing device, and actuators into one device and it can be attached to the user.
- In *distributed biofeedback systems* its elements are at the arbitrary positions, providing that they can perform their functions as intended. For example, sensors for motion acquisition can be attached to the user's body (accelerometer), or they can be located away from the user (video camera).

4.3.2 Functionality

Based on the intended *functionality* a number of different biofeedback system types can be defined. In this book we define the user, instructor, and cloud type of the biofeedback system.

- In the *user type* all the elements and functionalities of system are under user's control. For example, the user sets the intensity of the activity, monitored parameters, feedback information, feedback modality, etc.
- In the *instructor type* the instructor monitors and controls the biofeedback system operation. For example, the instructor analyses the performed action in post-processing, gives terminal feedback information to the user, and checks statistical parameters and history of the user.
- *Cloud type* system processes and/or stores the results in the cloud. Various types of data, from raw sensor signals to complex data analysis results, are available to anyone with appropriate access rights. Cloud data is usually accessed through a web application. Cloud system can also support a participatory concept of data acquisition, processing and sharing. For example, after years of collecting gait signals during the rehabilitation exercise, machine learning techniques are used to identify the most effective rehabilitation technique.

4.3.3 Physical Extent

In the terms of *physical extent* the biofeedback systems can be divided into personal, confined, and open space system types.

- In the *personal type* all system elements are in or on the user's body; for example a smartphone attached to user's lower back.
- *Confined space systems* have elements in the vicinity of the user; for example, in the same room or in the same playing field.
- *Open space systems* have no limitation of distances between the system elements; for example, alpine skiing, marathon run, cycling, and other open space sports.

In the latter two types, the element that is away from the user is in the great majority of cases the processing device. In some cases the sensing device, such as high-speed cameras can be distributed over the playing field or actuators, such as loudspeakers can be set somewhere close to the user. One of the limitations of open space systems is the latency of the communication channels that restricts the distance between system devices and the choice of communication channels.

4.4 Architectures

This section presents three examples of biofeedback systems based on different combinations of structure, functionality and physical extent, described in Sect. 4.3.1 to 4.3.2. We chose to name the different architectures based on their main functionalities; user, instructor, and cloud.

4.4.1 User Architecture

The architecture of a biofeedback system that belongs to the user functionality type is shown in Fig. 4.1. Its physical extend corresponds to the personal space type system, its structure resembles a compact type system. The monitoring device is used to control the operation of the biofeedback system and to view the results.

The user architecture can give either concurrent or terminal feedback. Because such systems are most commonly implemented on the embedded devices or on the smartphone, the computational power can be a problem, if the concurrent feedback is required.

An example of a user biofeedback system is a relatively undemanding application for gait control, where the entire system is implemented within a smartphone. Its sensors feed the application algorithm that concurrently outputs audio feedback signal correlated to one gait parameter, such as step symmetry or gait speed.

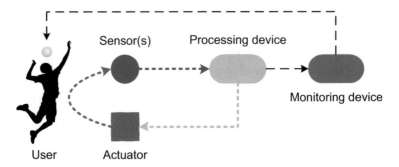

Fig. 4.1 User architecture of a biofeedback system. It resembles the user functionality type, its physical extend corresponds to the personal space type, and by structure it analogous to a compact type of the biofeedback system

4.4.2 Instructor Architecture

Biofeedback system shown in Fig. 4.2 belongs to the instructor functionality type, its physical extend corresponds to the confined space type, and by structure it resembles a distributed type of the biofeedback system. Its main difference to the user architecture is the distributed structure. In most cases the sensors and actuators are on the user, while processing and monitoring devices are with the instructor at the remote location.

This architecture requires wireless communication channels. If concurrent mode of operation is required, then the communication channels must have low latency and the processing device must be capable of performing all the necessary processing in real time.

An example of instructor architecture of the biofeedback system is a running monitoring application, where the instructor monitors the user's performance on a

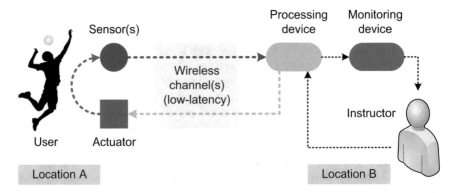

Fig. 4.2 Instructor architecture of a biofeedback system. It belongs to the instructor functionality type, its physical extend corresponds to the confined space type, and by structure it resembles a distributed type of the biofeedback system

stadium in real time. Processing and monitoring device can be, for example, a laptop, a tablet, or even a desktop. The feasibility of a concurrent feedback of this system in the major part depends on the properties of wireless communication channels.

4.4.3   Cloud Architecture

The architecture of a biofeedback system that belongs to the cloud functionality type is shown in Fig. 4.3. Its physical extent corresponds to the open space type, and by structure it resembles a distributed type of the biofeedback system. The core of the system is in the cloud that performs the functions of storage and data analysis, while the processing device is on the user or in its vicinity. When terminal feedback is sufficient, the processing device can also be in the cloud (not shown in Fig. 4.3).

Figure 4.3 indicates that the entire biofeedback loop can be at the user's location or in its vicinity. In the former case the loop operates in the user mode, in the latter case the loop operates in the instructor mode. In both cases loop signals (sensor and feedback) are sent to the cloud during the activity or after the activity for possible later complex data and statistical analysis. Results on a different level of detail are available to anybody with appropriate access rights—user, instructor or even general public. For example, users have access to the entire set of their own data, instructors

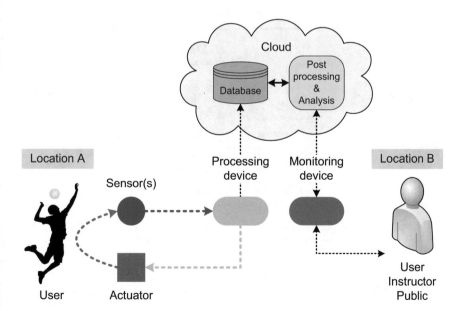

Fig. 4.3 Cloud architecture of a biofeedback system. It belongs to the cloud functionality type, its physical extend corresponds to the open space type, and by structure it resembles a distributed type of the biofeedback system

have access to most of the data of all users under their control, and general public has access only to anonymised and summarised data of all users or of a certain group of users.

An example of a cloud biofeedback system is an application that controls gait, but also send the gait signal to the cloud. In the cloud the data are processed, analysed and results are available to: (a) users to track their activity, (b) to instructors to give comments on performance and skill, and (c) to the public to see, for example, high level statistical data such as number of steps made each day.

4.5   Classification and Comparison

Biofeedback system architectures presented in Sect. 4.4 are of course not the only ones possible. Elements can be put together into a functional system in many different combinations according to the properties, demands, usage, constraints, and other influencing factors. For easier navigation among a number of possible combinations, Fig. 4.4 shows the classification of biofeedback systems architectures.

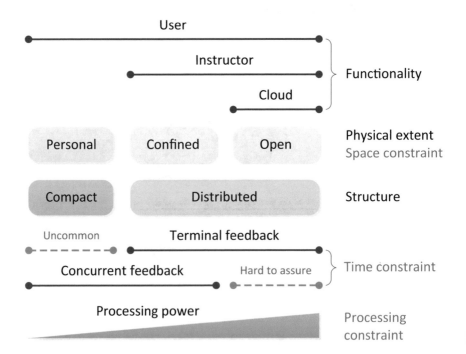

Fig. 4.4 Classification of biofeedback systems according to the properties listed in Sect. 4.3, constraints defined in Sect. 4.2, and architectures presented in Sect. 4.4 (Kos et al. 2018)

The classification is performed according to their different properties presented in Sect. 4.3 and different constraints defined in Sect. 4.2. For example, when the functionality is taken as the key aspect, the instructor architecture of the biofeedback system:

- can be devised as:

  - confined or open space system according to the space constraint or physical extent property;
  - distributed system in terms of structure;

- in the terms of the time constraint the system offers:

  - terminal feedback in confined and open space,
  - concurrent feedback in confined space;

- the required processing power ranges from moderate to high.

Many similar examples can be presented for other key aspect selections; for example, compact structure systems or confined space systems.

As already explained in Sect. 4.3.2, biofeedback system architectures are defined based on intended functionalities of the biofeedback system: user, instructor, and cloud. While the detailed comparison of the presented architectures based on all of the above presented aspects is out of the scope of this book, a direct comparison between individual architectures can be done based on Fig. 4.4. To assist the reader in this task, some of the most important advantages and disadvantages originating from the constraints described in Sect. 4.2 are presented.

4.5.1 User Architecture

User architecture has the most advantages, when implemented as a personal space system with a compact structure. All sensors, processing device and actuators are on the user's body, many times as a part of the same (embedded) system, i.e. smartphone. In general, sensor data synchronization and feedback loop delay are not a problem. Alternatively, to overcome the shortcomings of wireless transmission between the system elements, a wired communication medium can be used for data transfer.

It has the advantage of being an autonomous, lightweight, wearable system, usable in personal, confined, and open space.

The most notable disadvantage of such implementations is the limited processing power that can be an obstacle for implementation of concurrent feedback.

User architecture is applicable in many scenarios and situations. Some of the most obvious scenarios of use are self-training and recreation, suitable mostly for experienced users who are aware of their incorrect movements and know how to correct them.

4.5.2 Instructor Architecture

The strongest advantage of the *instructor architecture* is its flexibility and ability of overcoming the problems imposed by the given constraints. Since the processing device is away from the user it has no important limitations about the processing power needed for the implementation of concurrent feedback. It also allows the instructor to actively and timely participate in the user's activity.

The instructor plays an essential role in the operation of the system by advising the user on correctly performing the tasks and adjusting the parameters of the system. Depending on the function and the use of the biofeedback system, an instructor could be, for example, a coach, a trainer, a teacher, a therapist, a doctor, or a helpmate. Two basic instructor version subgroups are distinguished: confined and open space.

The main disadvantage of such implementations is the required use of wireless communication channels that may cause increased feedback loop delay or data loss. Especially in open space systems the transmission range or/and bit rate of the wireless technology can be a limitation factor.

One of the most obvious uses of the instructor architecture of the system is motor learning. In this scenario the instructor is monitoring the user's performance and progress through the biofeedback system analysis and visual contact. When necessary the instructor gives advice to the user or adjusts the parameters of the system to follow the learning progress of the user.

4.5.3 Cloud Architecture

The most notable advantage of the *cloud architecture* are its immense storage and processing capabilities that can be used for detailed analysis and data mining of the collected user activity data.

The main disadvantage of the cloud system at this time is the hard-to-control communication delay that makes concurrent feedback practically impossible.

The cloud architecture can be viewed as an addition or as an upgrade to all other biofeedback system architectures. In addition to the standard system devices (sensors, processing device, biofeedback device, and monitoring device), the cloud architecture includes storage and a range of cloud services that offer functionalities not available in any other architecture. One of the most straightforward services is a review of movements recently performed by the user. Other services include (but are by no means limited to) individual user movement comparisons, user/group analysis and comparison, user/group statistics, data mining, and tracking of user/group progress.

Location B in Fig. 4.3 is arbitrary, and the actions can be performed at any time after the data have been stored to the cloud and by anyone who has appropriate access to the cloud database. One of the many advantages of cloud approach is also the possibility of comprehensive interdisciplinary data analysis by various experts from different research fields.

Application scenarios of the cloud architecture include all possible scenarios from the user and instructor architectures; but with an important addition of the cloud functionality: post processing, second opinion, group statistics, time statistics, comparisons, etc. One such example is remote coaching.

References

Kos A, Milutinović V, Umek A (2018) Challenges in wireless communication for connected sensors and wearable devices used in sport biofeedback applications. Futur Gen Comput Syst

Kos A, Tomažič S, Umek A (2016) Suitability of smartphone inertial sensors for real-time biofeedback applications. Sensors 16(3):301

Umek A, Kos A (2016) Validation of smartphone gyroscopes for mobile biofeedback applications. Pers Ubiquit Comput 20(5):657–666

Umek A, Tomažič S, Kos A (2015) Wearable training system with real-time biofeedback and gesture user interface. Pers Ubiquit Comput 19(7):989–998

Chapter 5
Biofeedback Systems in Sport and Rehabilitation

5.1 Background

Biofeedback systems in sport and rehabilitation are particularly useful in motor learning (Giggins et al. 2013; Lauber and Keller 2014; Sigrist et al. 2013), as explained in Sect. 2.4. Recent advances in technology allow development of realistic and complex biofeedback systems with concurrent augmented feedback (Sigrist et al. 2013). Ubiquitous computing, with its synergetic use of sensing, communication, and computing is quickly entering sport applications (Baca et al. 2009). Sports applications are becoming increasingly mobile; the abundance of relatively inexpensive sensors is producing large amounts of data. Consequently information, communication, and computing technologies are becoming increasingly important in sport (Baca et al. 2009; Liebermann et al. 2002).

As discussed in Chap. 3, biomechanical biofeedback systems with augmented feedback use technical equipment for sensing, processing, and feedback, as shown in Fig. 3.1. The presentation of the current state-of-the-art technologies for biofeedback systems is given in Chap. 6.

In this chapter the focus is on the study of various parameters, elements, and properties of the biofeedback system that define its operation under different conditions and requirements specific for sport and rehabilitation. For example, general requirements of biofeedback applications in sport and rehabilitation are defined by position and/or orientation tolerance. The typical position errors allowed by biofeedback applications are up to a few centimetres, the typical angular errors are up to a few degrees. In golf biofeedback applications, the required accuracies are up to $2°$ in orientation and up to 1 cm in position (Umek et al. 2015). Sensors must exhibit sufficient accuracy, measurement range, and sampling rate to fulfil the above requirements and cover the biofeedback application movement dynamics.

© Springer Nature Switzerland AG 2018
A. Kos and A. Umek, *Biomechanical Biofeedback Systems and Applications*,
Human–Computer Interaction Series, https://doi.org/10.1007/978-3-319-91349-0_5

5.2 Sensing

The operation of biofeedback systems largely depends on parameters of human activity. Biofeedback in sport and rehabilitation is based on sensing human motion parameters: body rotation angles, posture orientation, forces, body translation, and body speed. Sensors measure the desired quantity or quantity that is in mathematical and/or physical relation to it. The most used are sensors based on the visual technology, motion measurement technology, force and pressure measurement technology, and muscle activity measurement technology. Important parameters of human motion should therefore be adequately acquired by a motion capture system with the following properties:

- sufficiently large dynamic ranges for the measured quantity,
- sufficiently high sampling frequency that covers all relevant frequencies contained in the motion,
- sufficiently high accuracy and/or precision.

Motion capture systems (MCS) are an important area of research connected to biofeedback systems. MCS employ various sensor technologies for motion acquisition. The most common are optical (camera) based systems and inertial sensor based systems (Liebermann et al. 2002). Optical systems generally give spatial positions of markers in the 3D space; inertial sensors generally give acceleration (accelerometer), angular speed (gyroscope), and orientation in space (magnetometer). Both of the commonly used MCS can be complemented by sensors measuring forces and pressures; the quantities that cannot be adequately measured by optical and inertial systems. Force, especially in some static positions, can be measured only by sensors for pressure (FlexiForce 2018) and strain gage sensors that measure strain, which is a relative deformation of the material when exposed to acting forces. Muscle activity and muscle tension are also very important quantities that can be measured by specially developed sensors (Đorđević et al. 2011).

MCS are bound by many requirements. For example, sensors on the athlete's body should not obstruct the movements, sensors in or on the sport equipment should not change its properties and functionalities.

5.2.1 Optical Motion Capture Systems

Optical camera based systems can be divided into two main subcategories: video based systems and marker based systems. The former directly process the video stream captured at various light wavelengths, the latter use passive or active markers for determining their position in space and time.

The most common are various video systems with terminal feedback (Schneider et al. 2015). Such systems usually record the exercise or a training episode, which is then replayed and visually analysed shortly after its execution. More sophisticated are systems with video processing that calculate the information about the desired movement execution parameters (Chambers et al. 2015; Schneider et al. 2015). In another group are motion tracking systems that use reflective markers to track body movement trajectories in 3D space. Examples of such systems are Vicon (Windolf et al. 2008) and Qualisys (Josefsson 2002; Motion Capture System 2018). Such systems may offer real-time functionality, but they are mainly designed to be used in closed and confined spaces with conditions favourable for visual marker tracking. They are also relatively expensive and demand a skilled professional to operate them.

5.2.2 Inertial Sensor Motion Capture Systems

The popular alternative motion tracking technology is based on inertial sensors. Measuring and quantification of human body activities and processes had grown very popular in recent years. A number of very different systems have been developed for that purpose.

Entry level systems and devices manly use only accelerometers to measure the energy expenditure of users. It has been confirmed that basic time-motion parameters in many sports can be analysed sufficiently well only by using 3D accelerometers. Most wearable devices include microelectromechanical systems (MEMS) inertial sensors (accelerometers, gyroscopes and magnetometers) integrated into one inertial measurement unit (IMU). Those inertial sensors are small, light-weight, and low-cost devices. For that reason, many inexpensive wristband fitness trackers are popular among general population (Seshadri et al. 2017; Li et al. 2016). Many wearable devices are commercially available and used in amateur and professional sport activities. A number of studies confirm that such devices correctly measure the number of steps and energy consumption, but are not reliable for more complex movement patterns or activities (Diaz et al. 2015; Takacs et al. 2014; Tucker et al. 2015). In addition to the physical body activity parameters, some wristbands and smart-watches also measure various physiological parameters, such as body temperature, heart rate, and blood oxygen level.

Beside the popular activity trackers for general population or for amateur users, other specialized devices are used in elite sports. They are able to accurately quantify specific athlete's activity and even reduce the number and severity of injuries. At the time of writing of this book, Catapult system has a leading role in elite sport performance analytics (The Monitoring System of Choice for Elite Sport 2018). Catapult wearable devices are designed to evaluate the sport-specific athlete's activity (player load) in more than thirty different sports, also in most popular team sports, such as football, rugby, and hockey. Catapult wearable devices are equipped with several sensors: accelerometer, gyroscope, magnetometer, and GPS. Xsens system (The Xsens wearable motion capture solutions 2018) uses multiple interconnected IMU devices including accelerometer, gyroscope, and magnetometer. Xsens system is aimed primarily at acquisition of accurate motion patterns of human body and its parts.

It is relatively straightforward to classify the abovementioned systems according to the rules presented in Sect. 4.5. Wristbands and smartwatches fall into the user functionality system type with compact structure and predominantly terminal feedback. The professional system Catapult falls into the instructor functionality system type with distributed structure and terminal feedback. All above described systems have elements of the cloud functionality because the data can be stored and processed in the cloud.

It should be emphasized that inertial sensor based motion tracking systems are generally mobile and have no limitation in covering space. Modern inertial sensors are miniature low-power chips integrated into wearable sensor devices. Their advantages are also their inexpensiveness, accessibility, portability, and ease of use. The drawbacks include possible high inaccuracies of the results when used without error compensation.

5.2.3 Sensor Properties and Limitations

To investigate the limitations of sensors used in biofeedback systems, the general properties and demands of biofeedback applications should be defined. Various parameters can be used for the evaluation of inertial sensors for a particular biofeedback application:

- Movement dynamics describes the swiftness of change in a movement, e.g., fast movements in sports and slow movements in rehabilitation biofeedback systems.
- In connection to movement dynamics, the required biofeedback system sampling frequency varies from a few tens to a few hundred Hertz.
- Dynamic range defines the boundary levels of the measured sensor signal (i.e., position in the 3D space, acceleration, and angular velocity).
- The duration of movement execution defines the width of the analysis time frame T_w that can vary from less than a second to a few minutes or even hours.
- The required accuracy of the measured or calculated movement parameters includes

  - The position accuracy acquired from the optical MCS or calculated from one or more inertial sensor readings; the maximal accumulated error ranges from a few millimetres to a few centimetres.
  - The posture angle accuracy, calculated from optical MCS data or accelerometer readings; the maximal accumulated error ranges from less than one degree to a few degrees.
  - The rotation angle accuracy, calculated from optical MCS data or gyroscope readings; the maximal accumulated error ranges from less than one degree to a few degrees.

- Measurement precision in biofeedback applications is many times more important than accuracy itself. When consistently repeating the same movement, measured values must be precise, even if they are not accurate.

One example of a biofeedback application in sports that can illustrate the application of the above defined biofeedback application parameters is the use of inertial sensors to measure the motion of the golf swing. It is a static exercise, but the body movements can be rather fast (high dynamic). Consequently, the sampling frequency must be high enough. Depending on the sensor placement, the accelerometer must cover different measurement ranges. For example, when attached to the lower part of the golf club, accelerations are much higher than when attached to the wrist or the arm of the player. The analysis time frame is between 2 and 3 s, depending on the player. The required accuracies are between 2 and 3 degrees for posture and angular rotation. The required measurement consistency is in the range of 2 degrees (Umek et al. 2015).

In the following two subsections, accuracy and sampling frequency, two of the most important properties of sensors, are presented through the experimental results using one optical MCS and one inertial sensor based MCS.

Sensor Accuracy

For the illustration of the sensor accuracy, we have experimentally evaluated two different motion capture systems: MEMS gyroscope based system and passive marker based optical system. We have focused on the accuracy of body rotations derived from the optical system spatial position measurements and MEMS gyroscope angular velocity measurements.

- *Optical motion capture system* - a professional optical motion capture system Qualisys™. This is a high-accuracy tracking system (Motion Capture System 2018) with eight Oqus 3 + high-speed cameras that offers real-time tracking of multiple marker points as well as tracking of pre-defined rigid bodies. Sampling frequency of the system is up to 1000 Hz. Accurate 3D position of each marker is obtained every millisecond. Infrared reflecting markers are attached to the rigid acryl frame, which is at the same time the encasement for the gyroscope. The markers are attached to the frame in a way to form the orthogonal vector basis of the rigid body in gyroscope's x-y plane.
- *Inertial sensors capture system* - MEMS gyroscope L3G4200D, manufactured by STMicroelectronics (ST Microelectronics 2010). Gyroscope device is fixed into the rigid acryl frame with the attached reflective markers of the optical system.

As stated in Guna et al. (2014) the measurement noise for a static marker of Qualisys optical system is given by its standard deviation for each individual coordinate: $stdx = 0.018$ mm, $stdy = 0.016$ mm and $stdz = 0.029$ mm. In the view of the given results, we can regard the measurement inaccuracy of the optical tracking system as negligibly small.

Figure 5.1 shows the comparison of the body Euler rotation angles measured by both tracking systems. A testing rotation pattern was generated by a smooth hand-driven object rotation sequence in a ten seconds time interval. The RMSE in the time

Fig. 5.1 Comparison of gyroscope (doted black) and Qualisys body rotation angles (solid colour) in the global sensor-body coordinate system. Red = roll, Green = pitch, Blue = yaw. The horizontal and vertical axes represent time [s] and angle [deg], respectively (Umek and Kos 2016)

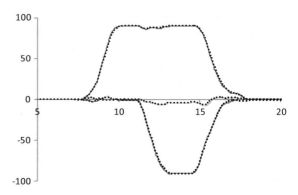

interval between 8 s and 18 s for the calculated rotation angles of the gyroscope are 1.11 deg, 0.81 deg, and 0.99 deg respectively. Such accuracies are good enough for most biofeedback systems (Umek et al. 2015).

It can be expected that similar results can be acquired by other MEMS gyroscopes of approximately the same quality.

Sampling Frequency

Due to real-time communication speed limitations of Qualisys, the above experiments are performed at sampling frequencies of 60 Hz (Nilsson 2011). While such sampling frequency is sufficient for evaluation of motion capture system accuracies that can be performed at low to medium movement dynamics, it is too low for capturing movements in sport. With the sampling frequency of 60 Hz only movements with maximal frequency component of 30 Hz or less can be captured correctly.

To estimate the required sampling frequencies for capturing human motion in sport, we performed a series of measurements with wearable Shimmer3™ inertial sensor device (Shimmer3 IMU Unit 2018). It allows accelerometer and gyroscope sampling frequencies of up to 2048 Hz. Maximal dynamic ranges of accelerometer and gyroscope are ± 16 g_0 (approx. 157 m/s^2) and ± 2000 deg/s respectively.

A set of time and frequency domain signals for a handball free-throw movement, sampled at 1024 Hz, is shown in Figs. 5.2, 5.3 and 5.4. The sensor device was attached to the dorsal side of the hand. The measured acceleration and rotation speed values are shown in Fig. 5.2. As it can be noticed, they are close to the limit of the sensors dynamic ranges set at 157 m/s^2 and ± 2000 deg/s.

High sampling rate of the inertial sensor device enables the measurements of actual spectrum bandwidth for both measured physical quantities. Most of the energy of finite time signals is within the upper limited frequency range. Figure 5.3 shows the accelerometer and gyroscope signal spectra between 0 and 60 Hz.

The bandwidth containing 99% of signal energy E(99%) is a useful measure of signal bandwidth as shown in Fig. 5.4. The signal spectra bandwidths differ in each dimension and are higher than that of absolute 3D values for each signal. The highest measured values in Fig. 5.4 are 59 Hz for acceleration and 40 Hz for rotation speed.

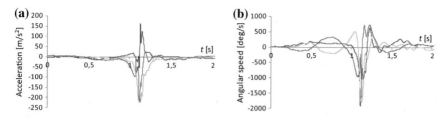

Fig. 5.2 An example of a high dynamic movement at a handball free-throw hand movement measured by Shimmer3 sensing device: **a** accelerometer and **b** gyroscope signals sampled at 1024 Hz (Umek and Kos 2016)

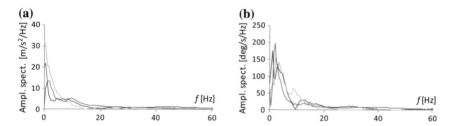

Fig. 5.3 An example of a high dynamic movement at a handball free-throw hand movement measured by Shimmer3 sensing device sampled at 1024 Hz. Signal spectrum (DFT) is calculated on the sequence of 2048 data points inside the 2 s time frame for **a** accelerometer and **b** gyroscope (Umek and Kos 2016)

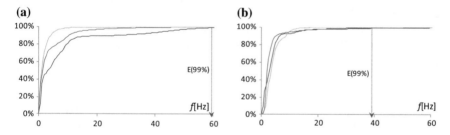

Fig. 5.4 An example of a high dynamic movement at a handball free-throw hand movement measured by Shimmer3 sensing device sampled at 1024 Hz. Signal bandwidth is measured and calculated by the relative cumulative energy criterion E(99%) for **a** accelerometer and **b** gyroscope (Umek and Kos 2016)

For some other, more dynamic, explosive movements we have measured the frequencies E(99%) that exceed 200 Hz, requiring sampling frequency of 500 Hz. All the experiments were performed by the amateurs and it is expected that professional athlete's movements are even more dynamic, requiring higher sampling frequencies, for example 1000 Hz. For the purpose of further discussion in this book we assume that the maximum required sampling frequency of biofeedback systems in sport is 1000 Hz.

5.3 Processing

The employed processing devices should have sufficient computational power for the chosen analysis algorithms. While this is generally not critical with terminal biofeedback that uses post-processing, it is of the outmost importance with concurrent biofeedback that requires real-time or and cyclic biofeedback that requires block processing. In biofeedback systems with real-time processing all computational operations must be completed within one sampling period and in biofeedback systems with block processing all computational operations must be completed within a short time period after each completed cycle. When sampling frequencies are high, these demands can be quite restricting, especially for local processing devices attached to the user.

5.3.1 System Implementations

In Chap. 4 a number of different biofeedback system architectures are studied, and compared. For the presentation and discussion of processing in biofeedback system in sport and rehabilitation, two basic biofeedback system implementations, according to Sect. 4.3.1, are selected based on the structure property: compact and distributed biofeedback systems that are shown in Figs. 5.5 and 5.6, respectively.

The *compact biofeedback system* is compact in the sense that all system devices are attached to the user and are consequently in close vicinity of each other, see Fig. 5.5. The system is performing *local processing*. Because the distances between devices are short, the communication can be performed through low-latency wireless channels or over wired connections. The primary concern of personal biofeedback

Fig. 5.5 Compact biofeedback system with local processing. All system elements are attached to the user. Wearable processing device tends to be the most critical element of the system in terms of its computational power and/or battery time

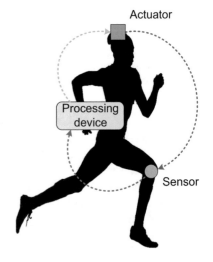

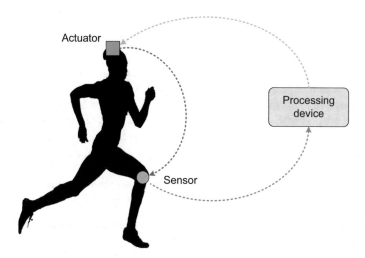

Fig. 5.6 Distributed biofeedback system. Sensor(s) and actuator(s) are attached to the user. The processing device is at the remote location, away from the user. Wireless communication channels tend to be the most critical element of the system in terms of range, bandwidth, and delays or any combination of the mentioned

systems is the available computational power of the processing device. The personal version is completely autonomous. The user is free to use the system at any time and at any place.

In the *distributed biofeedback system* sensors and actuators are attached to the user's body, while the processing device is at a remote location, see Fig. 5.6. The system is performing *remote processing*. Based on the distance between the user and the processing device, distributed systems are further divided into two subgroups: local and network. In the local subgroup the processing device is located close to the user. For example, just outside the playing field, next to the running track, at the base of the skiing slope, etc. In the network subgroup the processing device is located somewhere in the network. For example, on a PC in the laboratory or on the server in the cloud.

The primary concern of distributed biofeedback systems are communication channel ranges, bandwidths, and increased feedback loop delays. With the local subgroup of the system the user might be limited to a confined space if the communication channel technology has short coverage range. The advantage of distributed biofeedback systems, especially of the network subgroup, is their high computational power.

5.3.2   Motion Signal Processing

In wearable systems the battery time is often the most limiting factor; hence energy consumption in the biofeedback loop is of the prime concern. When having the possibility of using either a compact or a distributed version of the biofeedback system, one should consider choosing the system with the optimum energy consumption for the given task.

According to Poon et al. (2015) sensor devices consume many times more energy for radio transmission and memory storage than for local processing. This means that compact version with local processing could be more favourable option than distributed version with remote processing. In this context of the local processing it is implied that the signal is first (partially) processed inside the sensor device. The results are then communicated to the processing device of the biofeedback system for possible further processing.

Energy wise local processing at sensor device is very attractive, but there are some limitations that should be considered (Poon et al. 2015):

- Algorithms developed in wide spread software environments such as MATLAB are difficult to port to sensor devices.
- Sensor devices use microcontrollers for signal processing. They do not have a floating point unit and floating point operations must be simulated by using fixed point operation. This is slow and induces calculation errors.
- Total energy needed for all operations of one cycle could be higher than the energy needed for radio transmission of the raw data of the same cycle.
- Data from more than one sensor must be processed by a single algorithm instance.
- Computational load of the algorithm could be too high to be handled by the sensor device, i.e. the time needed to finish all operations of one cycle is longer than sampling time.
- More cooperative users are active in a biofeedback system at the same time.

When one or more of the abovementioned limitation apply, distributed system with remote processing is a better option. The advantages of the distributed biofeedback system are:

- The remote processing device has practically unlimited energy supply, high processing power, and large amounts of memory storage.
- The remote processing is flexible in the terms of software environments usage, algorithm changes, algorithm complexity, choice of technology, choice of computing paradigm, etc.
- Central point of sensor data synchronization when more users are simultaneously using a biofeedback system.
- High performance computing solutions can be used when the amounts of data and/or computational complexity increases.

The choice of the most appropriate biofeedback system depends on each separate application.

5.4 Communication

After the acquisition, the next step is getting the motion signals and data to the processing unit; and in the case of biofeedback systems also getting the feedback signals and data to the actuator. Motion capture systems can produce large quantities of sensor data that are transmitted through communication channels of a biofeedback system. When concurrent (real-time) transmission is required, the capture system forms a stream of data with data frames that are transmitted at every sampling episode, that is, with the frequency equal to the sampling frequency of sensors.

5.4.1 Transmission Delay

In Sect. 5.3.1 we distinguish two basic implementations of biofeedback systems; compact and distributed. In real-time biofeedback systems two main transmission parameters are important; bit rate and delay. While bit rate depends on the used technology, delay t_{delay} (Eq. 5.1) depends on signal propagation time t_{prop}, frame transmission time t_{tran}, and link layer or medium access control protocol (MAC) delay t_{MAC}.

$$t_{delay} = t_{prop} + t_{tran} + t_{MAC} \tag{5.1}$$

At a constant channel bit rate R, the transmission delay t_{tran} (Eq. 5.2) is linearly dependent on the frame length L.

$$t_{tran} = \frac{L}{R} \tag{5.2}$$

Propagation delays on different transmission media are 3.5–5 ns per meter. They are sufficiently small to be neglected. MAC delays vary considerably with channel load, from a few tens microseconds to seconds (rarely). In lightly to moderately loaded channel, MAC delays are normally below 1 ms. In most cases that leaves the transmission delay as the main delay factor in biofeedback systems. For example, according to Eq. 5.2, the transmission delay t_{tran} for a 100 bytes long frame transmitted over the link with a bit rate of 100 kbit/s is 8 ms. That is for a decade longer than the normal MAC delay.

5.4.2 Communication Technologies

Compact biofeedback systems can use body sensor network (BSN) technologies that have bit rates from a few tens of kilobits per second up to 10 Mbit/s (Poon et al. 2015). Considering the projected sampling frequency of 1000 Hz, that yields the maximal possible frame size in the range of a few tens of bits (a few bytes) for low-speed technologies and up to 10,000 bits (1250 bytes) for the high-speed technologies. The range of BSN is typically a few meters.

Distributed biofeedback systems use various wireless technologies with bit rates from a few hundreds of kbit/s up to few hundreds of Mbit/s (IEEE 802.11 standards 2018). Considering the projected sampling frequency of 1000 Hz, that yields the maximal possible frame size in the range of a few hundreds of bits up to 100,000 bits. The range of considered wireless technologies is between 100 m (WLAN technologies) and a few kilometres (3G/4G mobile technologies).

Communication is a problem in real-time biofeedback applications and in high-speed sensing in general. For example, sensor signals shown in Fig. 5.2 are acquired by logging and post-processing and not by streaming and processing in real-time. Although the employed Shimmer3 sensor device does support streaming, the bit rate of its available Bluetooth technology used for the transmission of sensor data is not high enough for streaming 6 DoF sensor signal data with sampling frequency of 1024 Hz.

It seems that wireless communication technologies are the most obvious choice for the transmission of data in biofeedback systems, but they are not the only possibility. Sensors can be connected to a sensor node or processing device by metal wires and optical fibres or, in the case of implants, use the human body as the propagation channel (Cavallari et al. 2014).

Various connectivity options and wireless technologies are shown in Fig. 5.7. As seen, sensors or sensor devices (IMU) can be connected to the remote processing unit (laptop, smartphone, cloud) directly or over a gateway device. The gateway synchronizes sensor signals and relays them to the processing unit. The gateway can also include some local processing capabilities.

The choice of the communication channels heavily depends on the type and dynamics of the sport's activity being monitored. Concerning the requirement for minimum obstruction of the user the most appropriate are systems with wireless communication.

Communication remains one of the urgent problems in sport feedback systems, especially in the case of concurrent biofeedback systems with real time operation. More detailed presentation and discussion of currently available communication technologies suitable for use in biofeedback systems in sport and rehabilitation can be found in Chap. 7.

5.5   Feedback

As explained in Sect. 2.3.1, biofeedback success is dependent on the proper feedback timing; therefore it is dependent on delays in the biofeedback loop. This is especially true and important for concurrent and cyclic biofeedback systems, but less for terminal biofeedback systems.

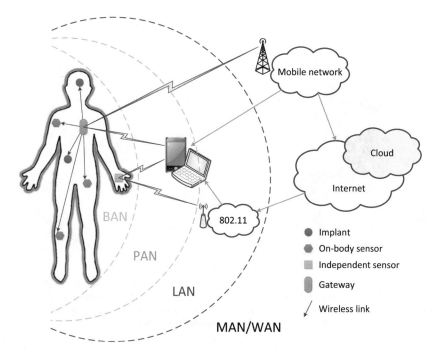

Fig. 5.7 Communication channels, technologies, and architectures that can be used in biofeedback systems in sport and rehabilitation. Sensors and sensor devices can be connected to the network directly or through the gateway. Processing can be done locally (gateway, sensor device) or remotely (smartphone, laptop, cloud)

5.5.1   Biofeedback Loop Delays

There are two basic points of view on delays in biofeedback systems: user's point of view and system's point of view. The delays in the biofeedback system are illustrated in Fig. 5.8.

From the user's perspective the feedback delay occurs during the user action; for example, during movement execution. In Fig. 5.8 this delay is depicted as *biofeedback delay*. This is the delay that occurs between the start of the user's action and the time the user reacts to the given feedback signal. For concurrent systems, biofeedback delay should be as low as possible. It is heavily dependent on user's reaction delay, which is not under the control of the biofeedback system devices and is dependent on human reaction time for the given task. To the best of our knowledge, there has not been any research that studies the time parameters of concurrent biofeedback in connection to human reaction times. For the purpose of this book, we define the maximal feedback delay to be a portion of the reaction delay, for example one tenth or one fifth of the reaction delay.

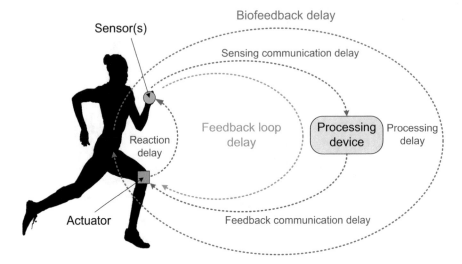

Fig. 5.8 Biofeedback system delays. Users perceive only the entire biofeedback delay (blue dotted line) defined as a sum of all delays in the system. The biofeedback system devices can control only the feedback loop delay, that is defined as a sum of all communication and processing delays of sensor(s), processing device(s), actuator(s), and communication channels

Reaction Delay

Reaction delay depends on human reaction times in different situations, different modalities, and different levels of task difficulty. Human reaction time is studied in numerous works. Results of reaction time measurements in (Jain et al. 2015; Human Reaction Time 1970; Pain and Hibbs 2007; Senel and Eroglu 2006) show that auditory reaction time (ART) is lower than visual reaction time (VRT). This is true for professional athletes, recreational sportsmen, and sedentary subjects.

The shortest reaction times are measured in sprint starts (Pain and Hibbs 2007). They can be below 100 ms, but sprint start is a very special case. Generally, trained athletes have ART between 150 ms and 180 ms, and VRT between 190 ms and 220 ms. Jain et al. (2015) have measured that the ART of sedentary and regularly exercising medical students is 229 ms and 219 ms respectively. Similarly, the VRT of both groups is 250 ms and 235 ms respectively. To cover the great majority of sports we should presume that the reaction time of a trained athlete is around 150 ms.

Different types of reaction times, dependent on different types of tasks are summarized in the below list.

- *Simple reaction time* is the time required for a subject to respond to the presence of a stimulus.

  - The subject might be asked to press a button as soon as a light or sound appears.
  - The mean reaction time for student population is about 160 ms to detect an auditory stimulus, and approximately 190 ms to detect a visual stimulus.

- The mean reaction times for sprinters at the Beijing Olympics were 166 ms for males and 189 ms for females, but in one out of thousand starts they achieved 109 ms and 121 ms, respectively.

• *Recognition or Go/No-Go reaction time* tasks require that a subject presses a button when one stimulus type appears and withhold a response when another stimulus type appears.

- The subject may have to press the button on a green light and not respond on a blue light.

• *Choice reaction time* tasks require distinct responses for each possible class of stimulus.

- The subject might be asked to press one button for a red light and a different button for a yellow light.

• *Discrimination reaction time* involves comparing pairs of simultaneously presented visual displays and then pressing one of two buttons according to which display appears brighter, longer, or greater in magnitude on something else.

Feedback Loop Delay

Human reaction times are a parameter that cannot be altered a lot and does not vary greatly among users. More interesting is the feedback loop delay that consists of system's devices processing delays and communication delays between them. The processing delays of sensors and actuators are considered to be negligibly low comparing to communication delays and processing delay of the processing unit; therefore, for the purpose of delay analysis in this book, we consider them to be zero. As defined above, feedback loop delay should be a small portion of the human reaction delay, which again depends on the modality (visual, auditory, haptic) used for the feedback.

Communication and processing delays within the feedback loop depend heavily on the parameters of the devices and technologies used. Some of the most important parameters are: sensor sampling frequency, processing unit computational power, communication channel throughput, and communication protocol delay. In general the feedback loop delay t_F is the sum of communication delay between the sensor(s) and the processing device t_{c1}, processing delay t_p, communication delay between the processing device and actuator(s) t_{c2}, sensor sampling time t_s, and actuator sampling time t_a:

$$t_F = t_s + t_{c1} + t_p + t_{c2} + t_a \tag{5.3}$$

In general the biofeedback system operates in the cycle that is equal to one of the system's device sampling time. Sensor sampling time is the most obvious choice. It should be emphasized that for the real time operation of the system it is required that the processing time of each sensor sample does not exceed the sensor sampling time $t_p \leq t_s$.

5.6 Real-Time Systems

Concurrent biofeedback systems with real-time augmented feedback are needed and required in many sports and rehabilitation tasks.

From the user's perspective, the feedback delay is the primary parameter defining the concurrency of a biofeedback system. In Sect. 5.5.1 we define that the feedback delay, that is the sum of all delays of the technical part of the biofeedback system (sensors, processing device, actuator, communication channels), should not exceed a small portion of the user's reaction delay. Also, as learnt in Sects. 5.3 and 5.4, processing and transmission delays are the most critical sources of delay in the feedback loop. Therefore we discuss their influences on the concurrent operation of the biofeedback system in some more detail.

5.6.1 Processing

Real-time processing is the most demanding example of human motion processing in biofeedback systems in the sense of required processing power of the processing device. In real-time biofeedback systems the processing device is receiving a stream of data frames with inter-arrival times that are averagely apart for system's sampling time T_s. To assure the real-time operation of the system, all operations on received data frame must be done within one sampling time; that is, before the arrival of the next frame.

The threshold of real-time operation of the processing device depends on many factors: computational power of the processing device, sampling time, amount of data in one streamed frame, number of algorithms to be performed on the data frame, complexity of algorithms, etc. It is therefore difficult to set exact thresholds or values for each parameter of the processing device.

Processing is a real problem in real-time biofeedback systems. For example, in Sect. 5.2 we present a comparison of optical and inertial sensor based capture systems that are operation in real time. In essence this comparison mimics the operation of a real-time biofeedback system to the point of the processing device. Despite the fact that Qualisys has video frame rates of up to 1000 Hz, the comparison could be done only up to sampling frequencies of 60 Hz. We identified the reason for this limitation in the processing load for the real-time calculation of the 6DoF orientation that could not be met by laptop processing power. It should be mentioned here that Qualisys by itself already is a high performance computing (HPC) system. It has eight cameras with integrated Linux system doing parallel processing of captured video. The results of marker positions are communicated to the central processing device (laptop) for synchronization and further processing.

The Need for HPC in Real-time Biofeedback

In Sect. 5.3.2 we have studied the trade-offs between the local and remote processing of biofeedback signals. While many examples of biofeedback applications exist, that do not require huge amounts of processing, one can easily find examples that require HPC.

One such example is a high performance real-time biofeedback system for a football match. Parameters at the capture side of the system are: 22 active players, 3 judges, 10 to 20 inertial sensors per person, 1000 Hz sampling rate, up to 13 DoF data. The data includes 3D accelerometer readings, 3D gyroscope readings, 3D magnetometer readings, GPS coordinates, and the time stamp. The first three sensors most often produce 16 bit values for each of the three axes, timestamp is 32 or 64 bit long, and GPS coordinates are 64 bits each. We must consider that GPS readings can be obtained only approximately 20 times per second. Taking the lower values of parameters (10 sensors, 32 bits for time stamp) the data rate produced is 44 Mbit/s. Taking the higher values of parameters (20 sensors, 64 bits for time stamp) the data rate produced is 104 Mbit/s.

5.6.2 Communication

Data rate values from the above example (44 Mbit/s and 104 Mbit/s) are calculated under the assumption that all sensor data is sent in binary format. Adding the protocol overhead, that is for example 30 bytes for IEEE 802.11 technologies, transmission rates on the communication channel are 104 Mbit/s and 224 Mbit/s respectively. Such data rates can be handled only by the most recent IEEE 802.11 technologies that promise bit rates in Gbit/s range.

Not all applications have such pretentious demands for processing and communication. For the illustration of the relations between the sampling frequency, bitrate of the communication channel, data size, and available processing time, two examples are given below. If the reaction time of trained athletes is defined at 150 ms (see Sect. 5.5.1) and that the maximal feedback delay is defined at 10% of user's reaction delay, then the maximal feedback loop delay must be less or equal to 15 ms.

- **Example 1:** Sampling frequency is 200 Hz (sampling time is 5 ms) and data size, including overhead, is 125 bytes per sensor sample and the same per actuator sample.

    - The remaining delay for the communication is 10 ms; 5 ms for the transmission of sensor samples and 5 ms for the transmission of samples for the actuator.
    - The required bitrate of the communication channel is approx. 200 kbit/s.

- **Example 2:** Sampling frequency is 100 Hz (sampling time is 10 ms) and data size, including overhead, is 500 bytes per sensor sample and the same per actuator sample.

– The remaining delay for the communication is 5 ms; 2.5 ms for the transmission of sensor samples and 2.5 ms for the transmission of samples for the actuator.
– The required bitrate of the communication channel is approx. 800 kbit/s.

The presented examples imply some form of high speed communication and some form of HPC, especially when complex algorithms and processes are used in combination with a large number of sensors. Algorithms that are regularly performed on a streamed sensor signals in biofeedback systems are (Ghasemzadeh et al. 2013; Lee et al. 2012; Min et al. 2010; Yurtman and Barshan 2014): statistical analysis, temporal signal parameters extraction, correlation, convolution, spectrum analysis, orientation calculation, matrix multiplication, etc. Processes include: motion tracking, time-frequency analysis, identification, classification, clustering, etc. Algorithms and processes can be applied in parallel or consecutively, depending on the algorithm flow.

References

Baca A, Dabnichki P, Heller M, Kornfeind P (2009) Ubiquitous computing in sports: a review and analysis. J Sports Sci 27(12):1335–1346
Cavallari R, Martelli F, Rosini R, Buratti C, Verdone R (2014) A survey on wireless body area networks: technologies and design challenges. IEEE Commun Surv Tutor 16(3):1635–1657
Chambers R, Gabbett TJ, Cole MH, Beard A (2015) The use of wearable microsensors to quantify sport-specific movements. Sports Med 45(7):1065–1081
Diaz KM, Krupka DJ, Chang MJ, Peacock J, Ma Y, Goldsmith J, … Davidson KW (2015). Fitbit®: an accurate and reliable device for wireless physical activity tracking. Int J Cardiol 185:138–140
Đorđević S, Stančin S, Meglič A, Milutinović V, Tomažič S (2011) Mc sensor - a novel method for measurement of muscle tension. Sensors 11(10):9411–9425
FlexiForce force sensors (2018) https://www.tekscan.com/product-group/embedded-sensing/force-sensors. Accessed 10 June 2018
Ghasemzadeh H, Ostadabbas S, Guenterberg E, Pantelopoulos A (2013) Wireless medical-embedded systems: A review of signal-processing techniques for classification. IEEE Sens J 13(2):423–437
Giggins OM, Persson UM, Caulfield B (2013) Biofeedback in rehabilitation. J Heuroengineering Rehabil 10(1):60
Guna J, Jakus G, Pogačnik M, Tomažič S, Sodnik J (2014) An analysis of the precision and reliability of the leap motion sensor and its suitability for static and dynamic tracking. Sensors 14(2):3702–3720
Human Reaction Time (1970–1979) The great soviet encyclopedia, 3rd edn. The Gale Group, Inc
IEEE 802.11 standards (2018) http://standards.ieee.org/about/get/802/802.11.html. Accessed 12 June 2018
Jain A, Bansal R, Kumar A, Singh KD (2015) A comparative study of visual and auditory reaction times on the basis of gender and physical activity levels of medical first year students. Int J Appl Basic Med Res 5(2):124
Josefsson T (2002) U.S. Patent No. 6,437,820. Washington, DC: U.S. Patent and Trademark Office
Lauber B, Keller M (2014) Improving motor performance: Selected aspects of augmented feedback in exercise and health. Eur J Sport Sci 14(1):36–43
Lee JB, Ohgi Y, James DA (2012) Sensor fusion: let's enhance the performance of performance enhancement. Proc Eng 34:795–800

Li RT, Kling SR, Salata MJ, Cupp SA, Sheehan J, Voos JE (2016) Wearable performance devices in sports medicine. Sports Health 8(1):74–78

Liebermann DG, Katz L, Hughes MD, Bartlett RM, McClements J, Franks IM (2002) Advances in the application of information technology to sport performance. J Sports Sci 20(10):755–769

Min JK, Choe B, Cho SB (2010) A selective template matching algorithm for short and intuitive gesture UI of accelerometer-builtin mobile phones. In: 2010 Second world congress on nature and biologically inspired computing (NaBIC), pp 660–665. IEEE

Motion Capture System (2018) http://www.qualisys.com. Accessed 10 June 2018

Nilsson L (2011) QTM Real-time Server Protocol Documentation Version 1.9. http://qualisys.github.io/rt-protocol/. Accessed 10 Sept 2015

Pain MT, Hibbs A (2007) Sprint starts and the minimum auditory reaction time. J Sports Sci 25(1):79–86

Poon CC, Lo BP, Yuce MR, Alomainy A, Hao Y (2015) Body sensor networks: In the era of big data and beyond. IEEE Rev Biomed Eng 8:4–16

Schneider J, Börner D, Van Rosmalen P, Specht M (2015) Augmenting the senses: a review on sensor-based learning support. Sensors 15(2):4097–4133

Senel O, Eroglu H (2006) Correlation between reaction time and speed in elite soccer players. Age 21:3–32

Seshadri DR, Drummond C, Craker J, Rowbottom JR, Voos JE (2017) Wearable devices for sports: new integrated technologies allow coaches, physicians, and trainers to better understand the physical demands of athletes in real time. IEEE Pulse 8(1):38–43

Shimmer3 IMU Unit (2018). http://www.shimmersensing.com/products/shimmer3-imu-sensor. Accessed 10 June 2018

Sigrist R, Rauter G, Riener R, Wolf P (2013) Augmented visual, auditory, haptic, and multimodal feedback in motor learning: a review. Psychon Bull Rev 20(1):21–53

ST Microelectronics. (2010). MEMS motion sensor: ultra-stable three-axis digital output gyroscope, L3G4200D Specifications. ST Microelectronics

Takacs J, Pollock CL, Guenther JR, Bahar M, Napier C, Hunt MA (2014) Validation of the Fitbit One activity monitor device during treadmill walking. J Sci Med Sport 17(5):496–500

The Monitoring System of Choice for Elite Sport (2018) https://www.catapultsports.com/products. Accessed 10 June 2018

The Xsens wearable motion capture solutions (2018) https://www.xsens.com/products/xsens-mvn/. Accessed 10 June 2018

Tucker WJ, Bhammar DM, Sawyer BJ, Buman MP, Gaesser GA (2015) Validity and reliability of Nike + Fuelband for estimating physical activity energy expenditure. BMC rts Sci Med Rehabil 7(1):14

Umek A, Kos A (2016) The role of high performance computing and communication for real-time biofeedback in sport. In: Mathematical problems in engineering, 2016

Umek A, Tomažič S, Kos A (2015) Wearable training system with real-time biofeedback and gesture user interface. Pers Ubiquit Comput 19(7):989–998

Windolf M, Götzen N, Morlock M (2008) Systematic accuracy and precision analysis of video motion capturing systems—exemplified on the Vicon-460 system. J Biomech 41(12):2776–2780

Yurtman A, Barshan B (2014) Automated evaluation of physical therapy exercises using multi-template dynamic time warping on wearable sensor signals. Comput Methods Programs Biomed 117(2):189–207

Chapter 6
Performance Limitations of Biofeedback System Technologies

6.1 Selected Technologies

Technologies used in biofeedback systems and applications are numerous and diverse; their detailed discussion or presentation is not in the scope of this book. However, some of those technologies are more important for the operation of the biofeedback systems and applications that are being designed than others. Likewise, some of those technologies present much greater obstacles to the developers than others.

We decided to thoroughly study and present some of the technologies that may be the most challenging in terms of their properties and limitations. The emphasis is on MEMS sensors and wireless communication technologies. For example, MEMS sensors' limitation is their accuracy that is not sufficient for some more demanding biofeedback applications; the main limitations of wireless communication technologies are their range and bit rate.

MEMS inertial sensors are studied through their properties in terms of their inaccuracies induced by biases and noise. Bias compensation options are presented and compensation strategies are proposed. Results of mass measurements of those parameters are carried out on inertial sensors integrated into smartphones. Bias compensation efficiency is presented through the inertial sensor performance comparison with the highly accurate professional optical system.

Wireless communication challenges are studied through the communication demands present in the biofeedback systems. They are most tightly connected to the requirements of data streams from sensors and to actuators as well as to the processing capabilities of the processing device. The most popular available and announced wireless technologies are presented and some guidelines for their selection are given.

© Springer Nature Switzerland AG 2018
A. Kos and A. Umek, *Biomechanical Biofeedback Systems and Applications*,
Human–Computer Interaction Series, https://doi.org/10.1007/978-3-319-91349-0_6

6.2   Requirements of Biofeedback Applications

Requirements of a wide variety of possible biofeedback applications are too diverse for a unified discussion. Since an important portion of this book is about biofeedback systems and applications in sport, requirements of sport applications in terms of technology are presented as an example.

Type of sport or a particular sport discipline is one of the major deciding factors for the correct choice of the most appropriate technologies used in development of the biofeedback systems and applications. The number of sport disciplines is far too large to be able to discuss the technical challenges in terms of biofeedback systems for each of them. Similarly to the classification of biofeedback system in sport, presented in Sect. 4.5, sport disciplines can be classified from the technical aspect based on the following major criteria:

- Place - fixed, bounded, unbounded.
- Number of users - individual, group, mass.
- Movement dynamics - low, medium, high.
- Movement type - defined single aperiodic movement, cyclic-periodic movement, not defined free movements.
- Equipment used - none, simple, complex.
- Environment - indoor, outdoor, water, etc.
- Analysis complexity - low, medium, high.

For example, biofeedback system for golf driving range training should be designed for a fixed place, one user, high movement dynamics, a defined single aperiodic movement, golf club is simple equipment, and the environment is indoor or outdoor, depending on the type of the driving range.

6.3   Inertial Sensor Properties

Inertial sensors used in sport predominantly fall into the group of *micro electrome-chanical systems* (MEMS). Inertial sensor errors have already been described well in various studies (El-Diasty et al. 2008; Aggarwal et al. 2008)) and can be classified into two groups: deterministic and random. Deterministic errors are bias, scale factor errors and axes misalignment errors. Random sensor errors are generated by various type of noise and depend on the sensor technology (El-Sheimy et al. 2008). There are other documented sources of MEMS errors (ST Microelectronics 2011; Sha-effer 2013) such as the effects of gravitational forces and vibrations. Deterministic errors can be greatly reduced by sensor calibration procedures, while noise reduction is limited and requires adequate noise filtering techniques. Noise reduction is also possible by using sensor arrays (Jiang et al. 2012). The most relevant determinis-tic error source for low-cost MEMS gyroscopes and accelerometers are their biases (El-Diasty et al. 2008).

Presently, thanks to smartphones, MEMS inertial sensors are readily available and widespread. All new smartphones are built with numerous sensors, practically always including accelerometers and gyroscopes. Because of the above factors, building biofeedback systems using the inertial sensors that are integrated into smartphones is desirable. This approach has many advantages: inertial sensors are already built into smartphones, representing a mobile wearable system with a powerful processing unit (CPU), a large battery, a high definition screen, many input/output interfaces, various choices of wireless connectivity, etc. Another advantage is the presumed synchronization of all sensor signals taken from the same smartphone. Less demanding biofeedback applications can be implemented entirely in the smartphone. Use of smartphones could also have some disadvantages; the most notable are their size and weight. When used as motion tracking sensors, smartphones (a) cannot be physically attached to certain parts and (b) could interfere with the movement being executed due to weight and size.

Extensive research has been conducted on the performances of standalone MEMS (Dixon-Warren 2010, 2011; El-Sheimy et al. 2008; Grewal 2010; Looney 2010; Shaeffer 2013; ST Microelectronics 2009, 2010, 2011; Stockwell 2004). There has been much less research specifically on smartphone-embedded MEMS. Measurements of standalone MEMS are generally done in controlled conditions, i.e., stable temperature, use of mechanical devices for accurate MEMS positioning, and at standstill (environment without vibrations). Smartphone-embedded MEMS rarely experience such controlled conditions. Therefore, it is essential that their performances are evaluated in normal operating conditions, i.e., with temperature changes due to the heating of other smartphone circuitry, measurements in slightly inaccurate positions, and in the presence of vibrations. Limited research about the latter issues can be found in Liu (2013).

The MEMS sensor parameters and resulting smartphone sensor parameters are not the only limiting factors for more demanding smartphone applications. Applications access MEMS sensor data and signals through the smartphone OS and APIs. One group of limitations is linked to the limited choice of available MEMS sensor parameters. For example, a MEMS accelerometer may have a set of different measurement ranges, e.g., ± 2 g, ± 4 g, and ± 8 g, but the operating system may only allow the use of ± 2 g. Another group of limitations is linked to the usage of smartphone resources, particularly its processing power. For example, the smartphone OS does not allow too frequent access to sensor data; therefore, the smartphone sensor data rate is limited to levels below the MEMS sensor data rates. APIs, software development tools, and libraries for smartphone applications can also be a problem. They may not include all of the necessary routines, protocols, and tools for building an effective application using smartphone sensors.

In this section, we focus on the performances, precisions, parameter measurement, bias determination and compensation of smartphone inertial sensors. We aim to determine the degree of usability of smartphone inertial sensors for real-time biofeedback applications. In the experiments, we use an iPhone 4 with an integrated 3-axis ST Microelectronics LIS331DLH accelerometer and a 3-axis ST Microelec-

Table 6.1 The main parameters of the iPhone 4 accelerometer and gyroscope (Kos et al. 2016a)

Parameter	3-axis accelerometer LIS331DLH	3-axis gyroscope L3G4200D
Range	±2 g–±8 g	±250–±2000 deg/s
Sensitivity	1±0.1 mg/dig	70 mdeg/s/dig
Bias error	±20 mg	±8 deg/s
Noise density	0.218 mg/sqrt(Hz)	0.03 deg/s/sqrt(Hz)
Sampling rate	0.5–1000 Hz	100/200/400/800 Hz

tronics L3G4200D gyroscope. Their main parameters of integrated sensors are listed in Table 6.1.

6.3.1 Accelerometer and Gyroscope Biases

Sensor bias is defined as an average sensor output at zero sensor input. Bias value in Eq. 6.1 is estimated by averaging N samples of sensor signal. The bias estimate averaging time depends on sampling frequency f_s and signal sample block length N. Bias estimate exhibits variations which are the result of a sensor noise:

$$x_{bias} = \frac{1}{N} \sum_{n=0}^{N-1} x[n] \tag{6.1}$$

Bias measurement results for accelerometers and gyroscopes from six smartphones of the same model are shown in Fig. 6.1. Measurements for all accelerometer and gyroscope axes were averaged over $N=600$ samples in the time interval $\tau = 10$ s.

The gyroscope biases are within the range of $\Delta G_0 = \pm 1.15$ deg/s. The accelerometer biases are within the range $\Delta A_0 = \pm 12$ mg$_0$ for the X- and Y-axes, and $\Delta A_0 = \pm 40$ mg$_0$ for the Z-axis. The aim of these results is to present bias variations on different smartphone devices of the same type.

6.3.2 Constant Bias Errors

The first evaluation of the measured biases in the scope of movement tracking and detection in biofeedback systems is done based on position and angle deviation from their real values. The largest measured accelerometer bias in Fig. 6.1 is 38 mg$_0$. The position drift Δs (Eq. 6.2) is a time quadratic function and consequently very sensitive to accelerometer constant bias Δa:

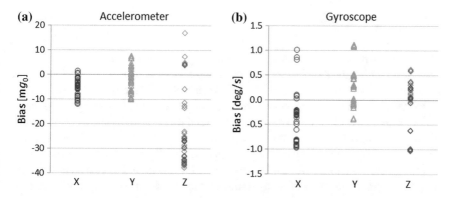

Fig. 6.1 3D gyroscope and 3D accelerometer biases obtained from multiple measurements on several smartphones. Values are shown in a scatter diagram with each of the axes plotted in a separate vertical column. Biases are calculated by averaging $N=600$ sensor signal samples at a sampling frequency $f_s = 60$ Hz; the corresponding averaging time is therefore $\tau = 10$ s. **a** Accelerometer biases presented in mg_0 have slightly different dynamic ranges. **b** Gyroscope biases presented in deg/s have similar dynamic ranges (Kos et al. 2016a)

$$\Delta s(t) = \frac{1}{2}\Delta a \cdot t^2 \tag{6.2}$$

The maximal position error from the measured accelerometer biases is 19 cm after 1 s and 19 m after 10 s. Therefore, accelerometer bias compensation is practically mandatory for movement tracking. The accelerometer signal can also be used to measure the direction of the gravity vector $\mathbf{g}_0$. Its direction error δ (Eq. 6.3) is proportional to the accelerometer bias vector $\Delta\mathbf{a}$:

$$\delta = \mathrm{acos}\left(\frac{\mathbf{g}_0 \cdot (\mathbf{g}_0 + \Delta\mathbf{a})}{|\mathbf{g}_0| \cdot |\mathbf{g}_0 + \Delta\mathbf{a}|}\right) \tag{6.3}$$

The predicted gravity vector direction error from the accelerometer measurement result in Fig. 6.1 is therefore less than 2.2 degrees. The constant gyroscope bias induces a linear angular drift $\Delta\Phi$ (Eq. 6.4):

$$\Delta\phi(t) = \Delta\omega \cdot t \tag{6.4}$$

The maximal measured linear angular drift for several different smartphones is approximately 1 deg/s. The gyroscope angle error is not too high. Uncompensated gyroscopes are potentially usable for short time movement analysis applications that do not require extreme precisions.

6.3.3 Bias Variation

MEMS accelerometer and gyroscope biases vary with time. One of the most well-known and influential factors in bias instability is temperature. Both sensor biases and scale factors are temperature sensitive parameters. For that reason, various MEMS techniques are used to compensate sensor temperature dependency (Aggarwal et al. 2006; Prikhodko et al. 2013). The thermal bias stability is also limited by temperature hysteresis error, which cannot be compensated (Weinberg 2011). Low-frequency bias changes are frequent due to partially uncompensated MEMS temperature variations.

Precise measurements of standalone sensor ICs (integrated circuits) can be done in temperature chambers with temperature regulation. Sensors in smartphones also experience changes in temperature because of other pieces of hardware integrated into the enclosed casing of the smartphone. At room temperature, the heating of the smartphone is predominantly caused by power dissipation from the CPU, screen, RF circuitry, GPS module, and other integrated circuits. Overheating of the batteries at charge time is also a possible cause. In its operating state, when applications are running and consuming power, the surrounding ambient air temperature has a smaller effect on the sensor temperature. Large internal temperature fluctuations are accompanied by large temperature-induced bias drifts in accelerometers and gyroscopes.

During the bias measurements in Fig. 6.1, we noticed differences in the biases of the same smartphone and between different smartphones. While differences between devices are due to variations in the physical properties of the embedded MEMS (Dixon-Warren 2010, 2011; ST Microelectronics 2009, 2010, 2011), the differences in successive measurements of the same device are caused by various inertial sensor instabilities, most likely because of different internal phone temperatures between measurements. To test our assumptions, we conducted a simple temperature stress test. First, we cooled the switched-off smartphone to 8°C. After turning the phone on and placing it in a location at room temperature (21°C), we measured the biases for three hours. During the tests, the smartphones were levelled using a levelling scale with an accuracy of 1 mm per meter or 0.057 degrees.

We performed measurements in different smartphone orientations with one of the axes parallel to the Earth's gravity vector. Examples of accelerometer and gyroscope bias drifts are shown in Fig. 6.2. The bias drifts in Fig. 6.2 exhibit the strongest temperature dependences in the first half hour of the test. In this period, the temperature changes induce bias drifts that can be larger than the bias variations caused by other sources of sensor error, including noise. The results of the stress test measurements of several smartphones show that bias drifts with similar dynamics are found on the other axes.

Temperature stress tests show that, as anticipated, the changes in temperature cause large bias variations. Stable temperature conditions are reached after approximately one hour. Biases measured after the transition period, do not change considerably. As shown in Fig. 6.2a, the variations in the accelerometer X-axis bias due to sensor noise do not exceed 0.4 mg_0, and bias drift after one hour reaches 0.7 mg_0.

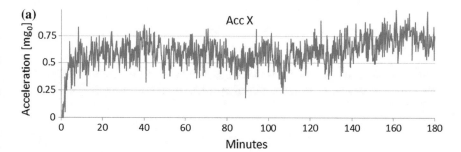

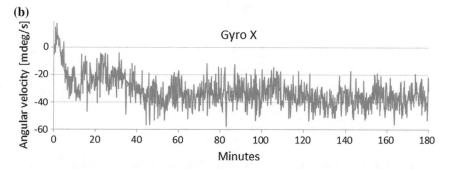

Fig. 6.2 Measured bias values in the 3-hour temperature stress test. The bias values are found by averaging 600 sensor signal samples in 10 s intervals. Temperature changes induce noticeable bias drifts in the accelerometers and gyroscopes. The graphs show the bias drifts of the **a** accelerometer X-axis and **b** gyroscope X-axis (Kos et al. 2016a)

The variations in the gyroscope X-axis bias are shown in Fig. 6.2b. The variations in the gyroscope bias due to sensor noise do not exceed 30 mdeg/s, and the gyroscope bias drift after one hour is approximately 40 mdeg/s.

Bias error measurements in stable ambient temperature conditions were carried out on six smartphones. The variations in the accelerometer bias due to noise are in the range of 0.4–0.5 mg_0. The variations in the gyroscope bias due to noise are in the range of 30–60 mdeg/s. Larger differences are found in the bias drifts. Measured accelerometer drifts do not exceed 4 mg_0 per hour, gyroscope drifts remain below 86 mdeg/s per hour.

6.3.4 Noise Measurement Methodology

For detailed analysis of the smartphone inertial sensor noise, we used Allan variance measurements. Biases are measured by averaging a finite sequence of samples when a device is in a standstill position. Bias variations are caused by various random processes in the operation of the sensor. Bias approximations $y[m]$ (Eq. 6.5) are calculated by averaging the sensor signal samples:

$$y[m] = \frac{1}{N} \sum_{n=0}^{N-1} x[n + m \cdot N] \tag{6.5}$$

Allan variance $\sigma_A^2(N)$ is a measure of the variations of the mean values $y[m]$ of consecutive blocks of N signal samples $x[n]$ (Allan 1966, El-Sheimy et al 2008; IEEE 1999):

$$\sigma_A^2[N] = \frac{1}{2}\overline{(y[m] - y[m-1])^2} \tag{6.6}$$

The variance is approximated from a finite number of mean values $y[m]$:

$$\sigma_A^2[N] \approx \frac{1}{2 \cdot (M-1)} \sum_{m=1}^{M-1} (y[m] - y[m-1])^2 \tag{6.7}$$

The approximation error of Eq. 6.7 is estimated to be (El-Sheimy et al. 2008)

$$\delta_\sigma = \frac{1}{\sqrt{2(M-1)}} \tag{6.8}$$

By definition Allan variance (Eq. 6.6) is a function of block length N, which can also be expressed by the block averaging time $\tau = N/f_s$, where f_s is the sampling frequency. Depending on the nature of the random process, the bias noise has a different power spectrum shape. In fact, many different uncorrelated random processes participate at the same time. The Allan variance method helps us to determine the characteristics of the underlying random processes and noise models. The basic bias error models are quantization noise, white noise, bias instability, rate random walk and drift rate ramp. Different noises have different spectrum power density profiles $S_x(f)$ and appear in Allan variance plots with different slopes (El-Sheimy et al. 2008; IEEE 1999):

- Quantization noise has an accented high frequency power density profile $S_x(f) = N_Q f^2$ and a quadratically decaying Allan variance $\sigma_A^2(\tau) = 3N_Q/\tau^2$.
- White noise, $S_x(f) = N_0$, has a linearly decaying variance $\sigma_A^2(\tau) = N_0/\tau$. Such noise causes *angle random walk* (ARW) in gyroscopes and *velocity random walk* (VRW) in accelerometers.
- Low-frequency noise has a power density function $S_x = N_B/f$ with a constant variance at long averaging time $\sigma_A^2(\tau) = 0.66 N_B$. It defines the minimum sensor *bias instability* (BI).
- *Random rate walk* (RRW) noise has a very long correlation time and, consequently, a narrower frequency power spectrum shape $S_x(f) = N_{RRW}/f^2$. Therefore, the variance is linearly proportional to integration time $\sigma_A^2(\tau) = 1/3 \, N_{RRW} \, \tau$.
- Some *deterministic errors* also causes slow monotonic bias changes, modelled as *drift rate ramp* (DRR) $x(t) = Rt$. The Allan variance function for this type of error is a quadratic function of averaging time $\sigma_A^2(\tau) = 0.5 \, R\tau^2$. For example, such bias

error is caused by temperature changes, and if not compensated, it can become a dominant error source in long time interval.

Various researchers have shown that, in most cases, different noises appear in different regions of τ. In such situations, it is possible to identify the model of the underlying random process from the Allan deviation $\sigma_A(\tau)$ log-log plot:

- High frequency variations caused by the quantization noise and white noise can be determined from the slope of the first segment of the Allan variance plot.
- Low frequency noise models start to dominate after filtering out the high frequency components by widening the averaging time τ.

Measurements were conducted under stable operation conditions when the smartphone was at standstill, i.e., absence of vibrations, constant room temperature, and constant low power dissipation. Information about the internal smartphone temperature or inertial sensor chip temperature was not available.

Measurements of Allan variance were carried out with the time resolution of one sample per octave from $\tau = 1/64$ s to $\tau = 1024$ s. For statistically valuable results, at least $M = 10$ measurements were required at the longest averaging time τ. Therefore, the measurement takes the time $T_0 = 10240$ s.

To determine all relevant noise terms, a finer resolution of the $\log(\tau)$ axis and longer measurement times are required. However, the results are still accurate enough to identify the most relevant types of bias errors: white noise and bias drift. Allan variance measurements of the 3D accelerometer and 3D gyroscope for a single smartphone are shown in Fig. 6.3.

As shown in Fig. 6.3a, the Allan deviation of the accelerometer $\sigma_A(\tau)$ follows the slope of the bias *white noise model* for short averaging times $\tau \leq 10$ s. The accelerometer *velocity random walk* constant (VRW) can be determined from the Allan deviation plot at $\tau = 1$ s: VRW$=\sigma_A(\tau = 1$ s). Model parameters for all three axes are given inside the shaded rectangle in Fig. 6.3a.

As shown in Fig. 6.3b, the Allan deviation of the gyroscope $\sigma_A(\tau)$ follows the slope of the bias *white noise model* for short averaging times $\tau < 100$ s. The gyroscope *angle random walk* constant (ARW) can be determined from the Allan deviation plot at $\tau = 1$ s: ARW$=\sigma_A(\tau = 1$ s). Model parameters for all three axes are given inside the shaded rectangle in Fig. 6.3b.

At longer averaging times, where the averaging filter decreases the power of the high frequency white noise, slow bias fluctuations with low frequency spectra become the dominant error sources for the accelerometers and gyroscopes.

The accuracy of the measured variances varies as estimated by Eq. 6.8. The Allan variance measurements of the most right side points, at $M = 10$ or $\tau = 1000$ s, are less accurate; the accuracy is estimated to be $\delta_\sigma = 23.6\%$ (Eq. 6.8).

The same measurements were performed on six different smartphones of the same type. We have noticed only minor differences in the white noise model parameters VRV and ARW. The differences in the bias instability were more noticeable, which we believe to be primarily the effect of the different temperature sensitivities of the smartphone embedded MEMS. The calculated average sensor white noise

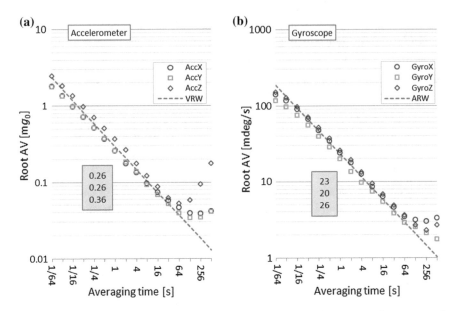

Fig. 6.3 Allan variance measurements for all three axes of the accelerometer and gyroscope of a single smartphone as a function of averaging time in seconds. The dotted line represents the white noise model: **a** Accelerometer results conform to the VRW model at short averaging times. **b** Gyroscope results conform to the ARW model at short averaging times. Values in rectangles represent Allan deviation at averaging time of 1 s (Kos et al. 2016a)

model parameters are VRW= 0.25 mg$_0$/$\sqrt{\text{Hz}}$ and ARW= 28 mdeg/s/$\sqrt{\text{Hz}}$. According to Dixon-Warren (2010) iPhone 4 embedded MEMS accelerometer is ST Microelectronics LIS331DLH MEMS and according to Dixon-Warren (2011) smartphone embedded MEMS gyroscope is ST Microelectronics L3G4200D. Their average noise model parameters are found in Table 6.1. It can be observed that noise parameters of the smartphone sensors offered by the operating system does not differ much from the same parameters of MEMS sensors.

6.3.5 Bias Measurement Error

Based on the measured sensor noise model, we can also determine the averaging times for the bias measurements that are used as the offset values for sensor bias compensation. Bias estimation error is a result of sensor noise passing through an averaging filter. The bias variance decays linearly with integration time only if white noise is a dominant source of error. Unfortunately, averaging cannot eliminate low frequency noise. The results from Fig. 6.3 help us analyse the trade-off between the signal averaging time and the bias measurement accuracy. Reasonable averaging times for bias measurements are between 10 and 100 s. The Allan variance (Eq. 6.7) can be used as an approximation of the standard bias variance (Allan Variance 2003; Hongwei et al. 2010; Land et al. 2007). The predicted bias measurement errors for the accelerometers and gyroscopes with defined confidence interval $k(P)$ are Eqs. 6.9 and 6.10:

$$\Delta a = k(P) \cdot \frac{VRW}{\sqrt{T_{avg}}} \tag{6.9}$$

$$\Delta \omega = k(P) \cdot \frac{ARW}{\sqrt{T_{avg}}} \tag{6.10}$$

The bias measurement errors induce drifts in position (Eq. 6.2), orientation (Eq. 6.3) and rotation (Eq. 6.4). Under the assumption that the sensor white noise is Gaussian (Leland 2005; Mohd-Yasin et al. 2003), $P(k=3)=99.7\%$ is the confidence interval. The bias variations due to the sensor noise shown in Fig. 6.2 are in accordance with the measured sensor noise density described by the ARW and VRW models from Eqs. 6.9 and 6.10, where bias variation peak-to-peak values in Fig. 6.2 are estimated to be within $\pm 3\sigma_A(T_{avg})$.

6.3.6 Influence of the Sensor White Noise on the Derived Parameters

Even with sensor bias compensation, sensor noise sets the lower bound of sensor precision. The Allan measurement results confirm that white noise is the dominant sensor error source for short integration times. Gyroscope white noise creates an angle random walk, where the standard deviation grows with the square root of the integration time (Woodman 2007):

$$\sigma_\phi(t) = ARW \cdot \sqrt{t} \tag{6.11}$$

Position random variations are the result of accelerometer signal double integration. Under the white noise accelerometer random model, the position variance, as a result of a second-order random walk, grows proportionally with $t^{3/2}$ (Woodman 2007):

$$\sigma_s(t) = \frac{VRW}{\sqrt{3}} \cdot t^{\frac{3}{2}} \tag{6.12}$$

We assume that the white noise in the accelerometer and gyroscope is Gaussian (Leland 2005; Mohd-Yasin et al. 2003), and therefore, both random variations Eqs. 6.11 and 6.12 represent 68% confidence intervals.

Table 6.2 Predicted random walk errors induced by accelerometer and gyroscope white noise. Values in the table represent standard deviations (Kos et al. 2016a)

Sensor	Error	Analysis time frame		
		1 s	3 s	10 s
Accelerometer	Position	0.14 cm	0.74 cm	4.48 cm
Gyroscope	Rotation angle	0.03 deg	0.05 deg	0.09 deg

Table 6.3 Predicted accelerometer position error in different bias compensation states and different analysis time frames

Bias error	T_{avg} (s)	Δa (mg$_0$)	Analysis time frame		
			1 s	3 s	10 s
A		40	19.60 cm	176.80 cm	1964.0 cm
B	10	0.24	0.10 cm	1.00 cm	11.7 cm
	100	0.08	0.04 cm	0.04 cm	3.7 cm
C - after one hour		4	2.00 cm	17.70 cm	196.4 cm

Table 6.4 Predicted accelerometer gravitation angle error

Bias error	T_{avg} (s)	Δa (mg$_0$)	Error (deg)
A		40	2.300
B	10	0.24	0.014
	100	0.08	0.004
C - after one hour		4	0.230

Table 6.5 Predicted gyroscope angle error in different bias compensation states and different analysis time frames

Bias error	Tavg (s)	$\Delta\Omega$ (mdeg/s)	Analysis time frame		
			1 s	3 s	10 s
A		1.15	1.15 deg	3.45 deg	11.50 deg
B	10	27	0.03 deg	0.08 deg	0.27 deg
	100	8.6	0.01 deg	0.03 deg	0.09 deg
C - after one hour		86	0.09 deg 0	0.26 deg	0.86 deg

Table 6.2 shows the predicted random walk errors of position and angle calculated by Eqs. 6.11 and 6.12. Values are given within a 68% confidence interval under assumption that sensor noise is Gaussian. For the majority of biofeedback applications, the random errors in Table 6.2, which are caused by noise, are negligibly small compared to the deterministic errors presented in Tables 6.3, 6.4 and 6.5 (Scenario A) and (Scenario C), which are caused by bias.

6.3.7 Bias Compensation Options

The precision of the sensor readings can be improved to a certain extent by bias compensation, but we have to bear in mind that bias errors can never be fully eliminated as shown and described in Fig. 6.4, which also illustrates various error components.

Uncompensated bias values that induce bias error A shown in Fig. 6.4b are presented in Tables 6.3, 6.4 and 6.5, which are based on the results of the example based on six smartphones presented in Sects. 6.3.1–6.3.4. From Table 6.5, we see that the largest gyroscope biases are approximately 1 deg/s, and from Table 6.3 that the largest accelerometer biases are approximately 40 mg_0. For the *uncompensated gyroscope*, in a short time analysis, for example, in a 3 s analysis interval, the bias induces an angular error of 3.45 degrees.

The precisions of the accelerometer and the gyroscope are considerably better after the bias compensation at time t_0 as shown in Fig. 6.4. The compensated bias values at time t_0 induce bias error B in Fig. 6.4b. The bias measurement error depends on the averaging time. The predicted bias measurement errors B from Tables 6.3, 6.4 and 6.5 represent 99.7% confidence intervals under the assumption that the sensor noise is Gaussian. For example, shortly after bias compensation, the expected linear

Fig. 6.4 Bias variations and their effects. **a** The bias changes with time (blue line) in the short term primarily because of bias noise (ARW) and in the long term because of other influences (red dashed line). Bias drift is the change in bias value between times t_0 and t_1. **b** Without compensation, we experience *bias error A*. With the compensation at time t_0, we decrease the bias error for the measured bias estimate to get *bias error B*. By time t_1, the bias drift causes the error to grow to the value of *bias error C* (Kos et al. 2016a)

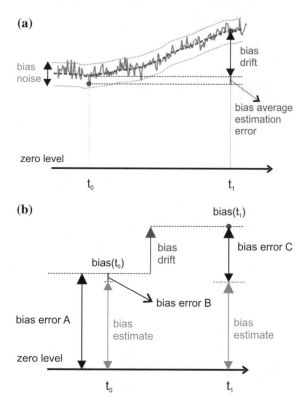

angular drift after a 3 s analysis time does not exceed 0.08 degree, and the expected positional drift remains under 1.0 cm.

If we perform the same analysis at time t_1, the expected angular drift would be higher because of the bias drift, which would result in bias error C, as shown in Fig. 6.4b. According to the results of the bias drift measurements from Sect. 3.3, the bias drifts are 4 mg_0 after one hour for the accelerometer and 86 mdeg/s after one hour for the gyroscope. For example, one hour after compensation, the bias drift induces positional error that does not exceed 17.7 cm and angular error that does not exceed 0.26 degrees in a 3 s analysis interval.

Tables 6.3, 6.4 and 6.5 show that the predicted errors one hour after compensation (C) are more than ten times smaller than those before initial compensation (A). If a new bias error induces angular and positional drifts that are no longer acceptable for the application, the bias drift should be corrected with another bias compensation at time t_1.

Based on the bias errors in Tables 6.3, 6.4 and 6.5 and the boundaries set for the exemplary biofeedback application presented in the introduction section, we can write the following findings:

Relatively high accelerometer biases restrict the use of uncompensated accelerometers for position tracking. Shortly after bias compensation, the accelerometer offers sufficient accuracy for applications with analysis times up to 3 s. Bias drift reduces the allowed analysis time to less than 1 s.

When an accelerometer is used for tilt or inclination sensing (static angles) in biofeedback applications, bias-induced errors are generally not the limiting factor. Static angle errors due to accelerometer noise and drift are practically negligible.

Bias Compensation Strategies

Inertial sensor bias variations in the form of noise and drift (see Fig. 6.4) could be the limiting factor for their usability in different types of applications. In biofeedback applications, where we generally use inertial sensors to measure movement patterns, large biases are a limiting factor.

With regard to each individual application and its sensor precision demands, we must choose the right strategy for bias compensation. There are several strategies available for bias compensation:

- *One-time bias compensation* has a time-limited effect. Therefore, it is suitable only for applications that operate in stable environments. For instance, the smartphone is always used inside the same temperature range and at approximately the same processing load.
- *Periodic bias compensation* can be performed at regular time intervals or on an as-needed basis. For instance, bias compensation is needed after every significant change in the inertial sensor temperature.

What are the available application scenarios according to the above options, according to the demands of the application, and according to the required times needed to perform the compensation? What about biofeedback applications that track and detect movements? To achieve different levels of movement detection accuracy, the following application scenarios are possible:

- The application uses uncompensated sensor data. In this case, the bias error corresponds to the bias error A in Fig. 6.4. This compensation scenario could be applicable to short time movements, up to a few seconds long, if the application does not use accelerometers for position calculation and does not demand high angular precision. See values for bias error A in Tables 6.3, 6.4 and 6.5.
- Before each session, the application performs a one-time bias compensation of the accelerometer and the gyroscope. Shortly after the compensation, the bias error corresponds to the bias error B in Fig. 6.4. With time, the bias error increases and corresponds to the bias error C in Fig. 6.4. The effect of one-time compensation is satisfactory up to one hour if the operating conditions do not differ much from the conditions at which the compensation was performed. In such cases, biases change in a limited value range, see values for bias error C in Tables 6.3, 6.4 and 6.5. Bias errors one hour after bias compensation are still for an order of magnitude lower than bias errors of uncompensated sensor. One-time bias compensation scenario is applicable to short time movements, up to a few seconds long, even for applications demanding high precision or for medium time movements, up to a few tens of seconds long for less demanding applications.
- The application constantly compensates biases. At every detected opportunity, the biases are compensated. The measurement times required for this compensation scenario could be between 10 s and 100 s, as proved in Sect. 3.4. After a longer time without compensation, the application may notify the user that the accuracy of the application operation might be compromised and that a new bias compensation is required. During the application use, gyroscope bias compensation is possible without too much trouble, while accelerometer bias compensation generally requires temporary interruption of application use (Stančin and Tomažič 2014). The goal in this scenario is to stay as close as possible to the accuracy level of bias error B in Tables 6.3, 6.4 and 6.5.

6.4 Smartphone Inertial Sensor Performance Comparison

Different biofeedback applications require differing degrees of sensor data quality; while some do not require highly accurate sensor data, others depend on it. The detailed inertial sensors properties can be found in manufacturers' data sheets. When sensors are integrated into IMU devices or smartphones, their data accuracy might change because of sensor signal processing within those devices.

A large share of MEMS inertial sensors are built into smartphones and a number of biofeedback applications are designed to run entirely on smartphones or use smartphone inertial sensors; similarly to the application presented in Sect. 7.3. The knowledge about available sensor accuracy provided by smartphones is therefore important for the developers of smartphone-based biofeedback applications (Kos et al. 2016b).

Similarly to the measurement sensor parameters of the six smartphones from Sect. 6.3, a wider measurement campaign has been carried out. Over 500 measurements of 116 different smartphone devices in 44 days have been collected. The collection of smartphones includes 61 different models of which 31 models are measured once and 30 models more than once. The measured smartphone models come from 13 different manufacturers and use two different platforms (Andriod, iOS). Measurement results are presented in two scales: (a) complete measurement results of all included smartphones and (b) results of individual smartphone devices through a number of measurements through time.

6.4.1 Measurement Results

The complete measurement dataset is presented by the measured devices. Plotted values represent the averaged measured parameter values by the device. Figure 6.5 shows smartphone accelerometer and gyroscope biases, Table 6.6 their average, standard deviation, and percentiles. Figure 6.6 shows smartphone accelerometer and gyroscope noise parameters VRW and ARW.

The averaged accelerometer biases for all devices under test are shown in Fig. 6.5a and averaged smartphone gyroscope bias measurements in Fig. 6.5b. Accelerometer and gyroscope biases should be measured and compensated in most sensor based applications. However, some less demanding applications can work even without bias compensation. The statistics on absolute bias values is collected in Table 6.6. Absolute bias values for more than 50% of measured devices stay under average and 95th percentile might be helpful for cross-platform application developers.

Accelerometer noise parameter VRW and gyroscope noise parameter ARW are obtained from single point Allan variation measurements under conditions defined in Sect. 6.3. Large deviations in measured values can be used to diagnose faulty devices. Figure 6.6 shows large deviations in measured noise parameters for both sensors, which are the result of different noise characteristics of MEMS sensor chips

Table 6.6 Statistical parameters of the measured smartphone accelerometer and gyroscope biases (Kos et al. 2016b)

	Accelerometer [mg_0]			Gyroscope [mrad/s]		
Parameter	X	Y	Z	X	Y	Z
Average	14.3	14.6	25.3	9.4	8.7	6.1
StDev	14.2	15.2	25.1	13.6	12.1	8.7
50th percentile	10.0	9.9	18.5	3.1	4.3	2.8
90th percentile	30.1	31.5	60.3	30.8	22.9	17.1
95th percentile	43.6	45.9	71.1	40.5	35.4	28.2
100th percentile	90.9	82.7	161.0	142.7	81.7	158.2

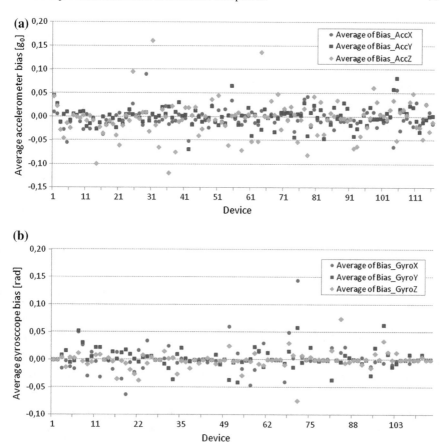

Fig. 6.5 Accelerometer and gyroscope bias measurements of X, Y and Z axes. **a** Average accelerometer biases of 116 different smartphones are plotted; **b** Average gyroscope biases of smartphones with gyroscopes are plotted. The horizontal axis represents the device identification number (ID) from the measurement database (Kos et al. 2016b)

embedded in different smartphone models. Noise parameter is further analysed after clustering the complete data set by smartphone model in the next section.

6.4.2   Long Term Bias Variation

A large number of bias measurements for the same sensor device can give some information about one of the most important sensor parameter; bias variation. Accelerometer and gyroscope biases vary with time. Bias variations are the result of random, low-frequency sensor noise and of deterministic dependence on temperature fluctuations. Deterministic bias drift cannot be compensated without measuring of sensor temperature.

(a)

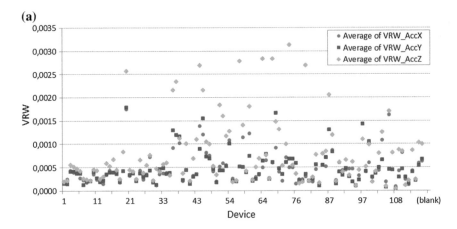

(b)

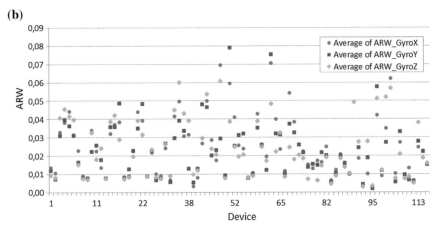

Fig. 6.6 Accelerometer (**a**) and gyroscope (**b**) noise of all properly functioning devices ($Ka = 108$, $Kg = 82$). The horizontal axis represents the device identification number from the measurement database. The vertical axis shows **a** VRW in $[g_0/\sqrt{Hz}]$ and **b** ARW in $[deg/s/\sqrt{Hz}]$ (Kos et al. 2016b)

Several smartphones have been repeatedly measured during the 44 days testing period. Figure 6.7 show the results of $N = 44$ simultaneous accelerometer and gyroscope bias measurements of the smartphone device with ID $= 3$. This device was under the test every evening repeatedly for 44 days. Bias variations pattern show that considerable variation is observed on a daily scale.

By closer inspection of measured sensor parameters of individual smartphones it can be seen that while biases vary between the smartphone models and within the same model, noise varies between the smartphone models and is stable within the same model.

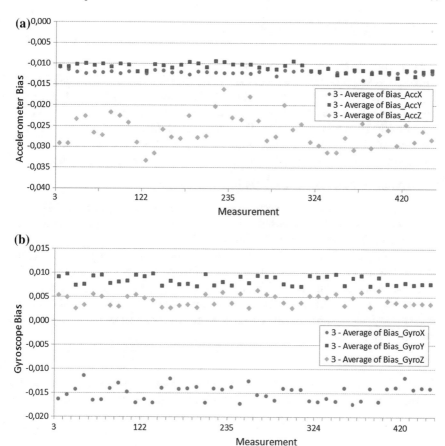

Fig. 6.7 Repetitive accelerometer and gyroscope bias measurements ($N = 44$) of the smartphone with ID $= 3$ showing bias variation. Measurement numbers are taken from the database and are not successive as other measurements took place in between two measurements of the presented device. Accelerometer bias is in [g_0], gyroscope bias is in [rad/s] (Kos et al. 2016b)

6.5 Motion Acquisition with Inertial Sensors

Smartphones are the most widely available mobile sensing devices. Thus, it would be highly beneficial if smartphone gyroscopes could have been used for biofeedback applications in sports, recreation, and rehabilitation.

In this section we use highly accurate optical motion capture system Qualisys (Qualisys 2018) for the validation of the smartphone gyroscopes. This and similar systems are regularly used as a *golden standard* in measurements in sport.

6.5.1 Experimental Design

The experiments are designed to track the angular motion of a 3D rigid body. Two different tracking systems are used: (a) a professional high-accuracy optical tracking system as a reference system and (b) a smartphone with an integrated MEMS gyroscope as the system for evaluation. The reference tracking systems measure the motion of the rigid body in the global Cartesian coordinate system, whereas the evaluated system measures the motion of the rigid body in the local coordinate system. The relation between both systems is explained in the following subsections.

Smartphone with an Integrated MEMS Gyroscope

The evaluated system is a smartphone-integrated MEMS gyroscope L3G4200D, manufactured by ST Microelectronics (2010). The detailed specifications of the gyroscope can be found in ST Microelectronics (2011).

The smartphone defines the local coordinate system and gyroscope rotation directions as shown in Fig. 6.8a. Rotation of the smartphone body around each of the local coordinate system axes yields the corresponding gyroscope signals, as shown in Fig. 6.8a. The x–y plane of the local coordinate system is parallel to the surface of the smartphone screen, and the z-axis is pointing upward when the phone is in the position with its face up.

Optical Tracking System

We used the optical motion capture system Qualisys (2018) as a reference for the 3D rigid body angular tracking. Qualisys is a professional, high-accuracy tracking

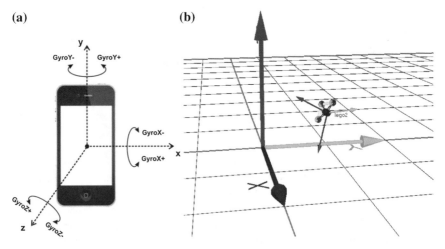

Fig. 6.8 a Smartphone with the definitions of the local coordinate system and gyroscope rotation directions. **b** Qualisys Track Manager 3D view window showing the position and orientation of the tracked rigid body. Reference frame (thick arrows) shows the global coordinate system. The body frame (narrow arrows) shows the local coordinate system (Umek and Kos 2016b)

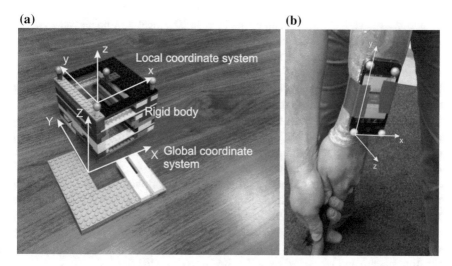

Fig. 6.9 Rigid body with the local and global coordinate systems in the test and golf movement setup (Umek and Kos 2016b)

system with eight Oqus 3 + high-speed cameras that offers real-time tracking of multiple reflective markers predefined rigid bodies. In view of the manufacturer stated parameters, we can regard the measurement inaccuracy of the reference system as negligibly small.

The motion of the rigid body is captured by the Qualisys Track Manager (QTM) software application and is displayed in a 3D view window, as shown in Fig. 6.8b. The QTM defines the global coordinate system, determines the 3D position of each tracked marker, and calculates the 3D orientation of the rigid body.

Technical Setup

Two sets of experiments with different setups were conducted. The first set of experiments comprises a series of hand-driven test movements of the custom-made rigid body. It is aimed at the precise measurement of the rigid body motion with well-defined positions at start, stop, and intermediate times. The second set of experiments comprises a series of golf swings. It is aimed at measuring the typical motion of the player's hand (rigid body) during the golf swing.

Lego bricks are used for the custom-made rigid body shown in Fig. 6.9a, which is also the encasement for the smartphone. It is possible to find Lego bricks that allow a perfect fit for the smartphone in the Lego frame. Lego bricks are used because they are widely available, offer high adaptability, and are manufactured with an accuracy of 10^{-5} m.

Four infrared reflecting markers are attached to the rigid body, and the smartphone is tightly embedded into the Lego frame. Three markers are attached to the frame to form the orthogonal vector basis of the x–y plane and define the local coordinate system xyz of the rigid body, as shown in Fig. 6.9a. The local coordinate system of the rigid body is aligned to the smartphone's coordinate system, as shown in Fig. 6.8a.

The series of hand-driven test movements is performed on a stable, levelled wooden table. The origin of the global coordinate system XYZ is defined by the Qualisys reference motion tracking system and is marked with the Lego plates that are firmly attached (glued) to the table, as shown in Fig. 6.9a.

For the golf swing movement, the smartphone is attached directly onto the forearm of the player, as shown in Fig. 6.9b. Four infrared reflecting markers are attached directly to the smartphone, three of which are attached to form the orthogonal vector basis of the x–y plane of the local coordinate system of the rigid body. The local coordinate system of the rigid body is aligned to the smartphone's coordinate system xyz shown in Fig. 6.8a. The origin of the global coordinate system XYZ defined by the reference system is not visible in the picture.

The series of golf swing movements is performed on the laboratory floor. The origin of the global coordinate system is defined by the Qualisys reference motion tracking system and is marked by self-adhesive tape on the floor.

Methodology

The smartphone gyroscope is evaluated through the comparison of rigid body orientations gained from the smartphone gyroscope and QTM signals.

The Qualisys 6DOF tracking function computes the body origin vector $\mathbf{P}_{origin}$ and the rotation matrix $\mathbf{R}$, which describes the rotation of the rigid body, as illustrated in Fig. 6.8b. Both parameters uniquely define the current position and orientation of all rigid body points $\mathbf{P}_{local}$ in the global coordinate system $\mathbf{P}_{global}$:

$$\mathbf{P}_{global} = \mathbf{R} \cdot \mathbf{P}_{local} + \mathbf{P}_{origin} \qquad (6.13)$$

Only the rotation matrix or equivalent Euler angles (roll, pitch, and yaw) are needed for the gyroscope evaluation. The real-time motion-tracking data stream originating from the Qualisys Track Manager software is captured by the Qualisys LabVIEW client. The gyroscope data stream from smartphone is captured by the custom-designed LabVIEW application. Both data streams are synchronized inside the main LabVIEW signal processing loop running at 60 Hz. The reference QTM system and evaluated gyroscope systems cannot be directly compared for two reasons: QTM gives rotation angle data (roll, pitch, and yaw), and the gyroscope gives angular velocity data. The aforementioned physical quantities are expressed in two different coordinate systems (local and global).

System Comparison Methods

We identified two methods for comparing results from the smartphone gyroscope and the optical tracking system.

Transformation of QTM rotation matrices $\mathbf{R}_{QTM}[n]$ (Eq. 6.14) to partial rotation matrices of the sensor-attached rigid body $\mathbf{R}_{local}[n]$ (Eq. 6.15) and calculation of the corresponding body rotation angles $\Delta\mathbf{\Theta}_{local}[n]$ (Eq. 6.16) between successive QTM analysis frames, thus enabling the evaluation of the *virtual gyroscope* data

$\boldsymbol{\Omega}_{\mathbf{QTM}}[n]$ (Eq. 6.17). The parameter f_s represents the synchronized sampling rate. The comparison of the reference system with the evaluated system is expressed in the local sensor-attached body coordinate system by the angular error $\boldsymbol{\varepsilon}_{\text{local}}[n]$ (Eq. 6.18).

$$\mathbf{R}_{QTM}[n] = \mathbf{R}_{global}[n] = \prod_{i=1}^{n} \mathbf{R}_{local}[i] \tag{6.14}$$

$$\mathbf{R}_{local}[n] = \mathbf{R}_{global}^{-1}[n-1] \cdot \mathbf{R}_{global}[n] \tag{6.15}$$

$$\Delta\boldsymbol{\Theta}_{local}[n] = \boldsymbol{\Theta}(\mathbf{R}_{local}[n]) \tag{6.16}$$

$$\boldsymbol{\Omega}_{QTM}[n] = \Delta\boldsymbol{\Theta}_{local}[n] \cdot f_s \tag{6.17}$$

$$\boldsymbol{\varepsilon}_{local}[n] = T_s \cdot \sum_{i=1}^{n} (\boldsymbol{\Omega}_{QTM}[i] - \boldsymbol{\Omega}_{gyro}[i]) \tag{6.18}$$

Gyroscope data $\boldsymbol{\Omega}_{\text{gyro}}[n]$ are used to calculate the successive local body rotation angle vectors $\Delta\boldsymbol{\Theta}_{\text{local}}[n]$ (Eq. 6.19), where T_s represents the sampling time. Transformation of successive rotation matrices of the sensor-attached marked rigid body $\mathbf{R}_{\text{local}}[n]$ (Eq. 6.20) to the global coordinate system rotation matrices $\mathbf{R}_{\text{global}}[n]$ (Eq. 6.21), followed by the calculation of equivalent Euler angles around all principal axes $\Delta\boldsymbol{\Theta}_{\text{global}}[n]$ (Eq. 6.22) (roll, pitch, yaw). The angular error $\boldsymbol{\varepsilon}_{\text{global}}[n]$ is expressed in the global reference coordinate system (Eq. 6.23), where $\boldsymbol{\Theta}_{\text{QTM}}$ represents the QTM body Euler angle vector.

$$\Delta\boldsymbol{\Theta}_{local}[n] = \boldsymbol{\Omega}_{gyro}[n] \cdot T_s \tag{6.19}$$

$$\mathbf{R}_{local}[n] = \mathbf{R}(\Delta\boldsymbol{\Theta}_{local}[n]) \tag{6.20}$$

$$\mathbf{R}_{global}[n] = \mathbf{R}_{global}[n-1] \cdot \mathbf{R}_{local}[n] \tag{6.21}$$

$$\boldsymbol{\Theta}_{global}[n] = \boldsymbol{\Theta}(\mathbf{R}_{global}[n]) \tag{6.22}$$

$$\boldsymbol{\varepsilon}_{global}[n] = \boldsymbol{\Theta}_{QTM}[n] - \boldsymbol{\Theta}_{global}[n] \tag{6.23}$$

The transformation formulas between rotational matrices $\mathbf{R}(\boldsymbol{\Theta})$ and Euler angles $\boldsymbol{\Theta}(\mathbf{R})$ are expressed by the rotation sequence around all three axes in a defined order (x, y, z), which is a default convention in QTM. The QTM reference system and evaluated system with smartphone MEMS gyroscopes are compared in both coordinate systems in Sect. 6.5.2.

6.5.2   System Comparison and Validation

The smartphone gyroscope evaluation procedure is based on measurements of rigid-body angular motion. The measurements are conducted concurrently in both systems: the reference professional optical tracking system and the evaluated smartphone gyroscope. Angular motion is expressed and compared by one of the two proposed

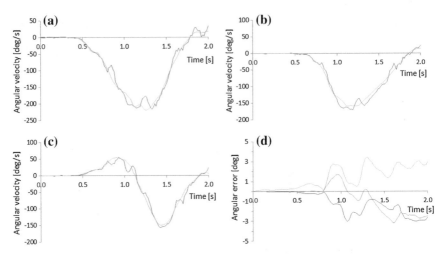

Fig. 6.10 Comparison of smartphone gyroscope signals (blue plots) and the derived reference system virtual gyroscope signals (red plots) in the local coordinate system. Graphs **a–d** show signals of the *golf swing movement*. Graph **a** shows the smartphone and virtual gyroscope x-axis (roll) angular velocity, graph **b** shows the smartphone and virtual gyroscope y-axis (pitch) angular velocity, and graph **c** shows the smartphone and virtual gyroscope z-axis (yaw) angular velocity. The difference in orientation angles is shown in graph **d** (colour code: red = roll, green = pitch, blue = yaw) (Umek and Kos 2016b)

methods presented and explained in Sect. 6.5.1 (System comparison methods). The sampling rate of both systems is set at 60 samples per second.

The golf swing is executed in full, but only its backswing component is tracked. The backswing phase measurement takes approximately 2 s, with 1.5 s of observed movement.

The evaluated gyroscope data are compared with reference virtual gyroscope data using Eqs. 6.14–6.18, and the results are presented by two sets of measurements. Figure 6.10 shows the comparison results of measured 3D rotation angles presented in the local coordinate system. Graphs (a)–(c) show golf swing measurements. Some high-frequency noise can be observed in all three virtual gyroscope signal components. However, when calculating angular orientation error (Eq. 6.18), high-frequency noise components are largely filtered out. The orientation angle error in the local coordinate system for the golf swing is shown in graph (d); the calculated RMSE is 1.86°.

From the experiment observer's view, it is more natural and more convenient to express the rotation angles in the global coordinate system. Rotation angles calculated from gyroscope data using equations Eqs. 6.19–6.23 are compared with the reference QTM rotation angles in Fig. 6.11. Graph (a) shows the rotation angles given by both systems, and graph (b) shows the rotation angle errors in the global coordinate system (Eq. 6.23); the calculated RMSE is 1.15°.

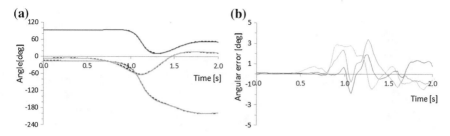

Fig. 6.11 Comparison of smartphone (dotted black plots) and QTM body rotation angles (solid coloured plots) in the global coordinate system (colour code: red = roll, green = pitch, blue = yaw). Graph **a** shows smartphone and QTM rotation angles of *golf swing movement*. Graph **b** shows the difference in rotation angles of the QTM system (reference) and smartphone gyroscope (Umek and Kos 2016b)

The comparison results in Figs. 6.10 and 6.11 are obtained after employing the gyroscope bias compensation and gyroscope scaling factor calibration.

Smartphone Gyroscope Validation

Based on the comparison results presented in this section, smartphone gyroscopes are validated for angular motion tracking in mobile biofeedback applications. The typical required angular accuracies of biofeedback applications are up to a few degrees; after full gyroscope calibration, the measured inaccuracies are 0.42° for the test movement and 1.15° for the golf swing; under partial gyroscope calibration, they are 2.05° for the test movement and 1.67° for the golf swing (bias compensation only). The measurement results confirm that the gyroscope precision is adequate for most biofeedback applications.

6.6 Processing and Communication

The need for processing and communication resources of the biofeedback system depends on a number of factors; from the number of sensors, their sampling frequency and bit rate to the available battery power, distance between system element, available communication technologies, protocol stacks, and other.

Processing power and processing delay are especially important in real-time biofeedback systems where the processing device is receiving streamed sensor data at every sensor sample period. Communication technologies should have high enough bit rate to support the transmission of sensor data to the processing device. The processing of received sensor data must be done within one sampling period. In a pipelined processing algorithm the processing delay can be several sampling periods, but a new processing result must be available at every sampling period. A binary classification of processing devices is possible into (a) devices capable of real-time processing and (b) devices capable of post processing. In connection to the process-

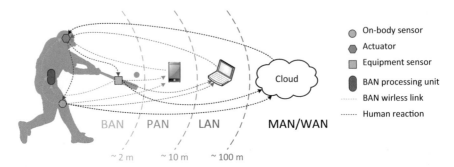

Fig. 6.12 Sensors, communication and processing resources in the biofeedback loop (Kos et al. 2018)

ing requirements, the bit rate of communication channels is of limited importance, providing that it is high enough to assure the transmission of all data created by sensors. More about related issues can be found in Umek and Kos (2016a).

An important factor in processing and communication demands of real-time biofeedback systems is the number of sensors in the system. The processing power and bit rate of communication technology of multi-sensor systems must be proportional to the number of sensors. Apart from the increased processing and communication demands, other problems can become an issue. For example, sensor data stream synchronization or sensor node density. The former is important for a quality sensor data processing, the latter is important for the correct selection of communication technology, which must allow concurrent operation of a large enough number of nodes. Some more details about these issues can be found in Sect. 7.1.2.

Some of the many possible combinations of biofeedback system architectures presented in Sect. 4.4 can be explained by Fig. 6.12. For the connection with wireless technologies presented in Sect. 6.6.3, networks classification by area is included.

A personal space biofeedback system can be realized within the Body Area Network (BAN), where body attached sensors and actuators use wired or wireless communication channel, and equipment sensors use wireless communication channel. The functionality roughly corresponds to the user architecture of the biofeedback system and the BAN processing unit collects data for post processing or performs real-time processing for concurrent feedback.

A confined space biofeedback system with the instructor functionality can be easily realized within the Local Area Network (LAN) or within the Personal Area Network (PAN). The sensors and actuators are wirelessly connected to a laptop or a notepad over the wireless LAN technology (WLAN). Similarly to the previous example the LAN processing unit performs post processing or real-time processing.

An open space personal biofeedback system can be realized as a cloud application. Body attached sensors, sport equipment sensors and actuators communicate with the cloud over the Wide Area Network (WAN), or they are wirelessly connected to a gateway that does the WAN communication with the cloud. Real-time operation is made difficult due to larger delays in communication with the cloud. Post processing options are practically unlimited.

6.6.1 Signal and Data Processing

Signals and data processing in sport feedback systems ranges from relatively undemanding to extremely demanding and time consuming. The processing needs on one hand and the processing capabilities on the other hand depend on a number of factors and situations: time of processing, place of processing, processing complexity, available processing power, available battery capacity, etc.

The time of processing depends on the type of feedback. If the feedback is concurrent, given during the action, the processing must be performed in real time. If the feedback is terminal, given after the action is completed, then the system can afford to do everything in post-processing.

The place of processing can be local, near-local, and remote. In the local case all of the processing is performed by the sensor device or by the gateway. Two main possible problems of local processing are available local processing power and locally available energy (battery). Local processing is performed by embedded devices; it is suitable and convenient primarily for low complexity real time biofeedback systems. Near-local processing is performed relatively close to the action. The main possible problems are the limitations of short-range communication technologies, especially in the case of concurrent biofeedback systems. Processing power can be a problem with the use of smartphones, less likely with the use of a laptop or a personal computer. Remote processing is done by any device connected to the Internet, most probably in the cloud or supercomputing centre. The main possible problem is the limitation of long-range communication technologies, especially their latency.

Complexity of processing depends on the amount and dimension of data, sampling frequency, algorithms, required precision, data analysis methods, etc. Processing can range from simple event counting to machine learning and data mining. Techniques and methods used depend on the intended results and timing urgency. With terminal feedback systems we can afford long processing times with complex analysis methods, but with concurrent biofeedback systems feedback delay is the most important factor that limits our possibilities. In concurrent biofeedback systems we cannot afford complex data analysis, we can mainly rely on signal processing algorithms and statistical methods. For example, we can trigger the feedback on a pre-set signal threshold or we can track statistical measures of the current signal and trigger feedback when their deviation is too high. In terminal feedback there is much less limitation. For example, we can collect motion data for longer time periods and then apply machine learning algorithms that classify the collected movements into several groups according to the movement execution quality. This allows for almost unlimited options for data processing.

Real-Time Processing Challenge

Sensor sampling frequency should be adjusted according to the movement dynamics, measured by the highest frequency in the sensed signal (position, angular velocity or acceleration) acquired by wearable sensor devices attached to the user. Some measured values indicate, that the highest frequencies in the sensed movement are: (a) up to 20 Hz for low dynamic movement, (b) up to 50 Hz for medium dynamic movements, and (c) up to 70 Hz for high dynamic movement. Much higher sampling frequencies, up to 1000 Hz, are needed when sensors are integrated into sport equipment.

Sensor sampling frequency also defines the lower limit of the feedback loop delay, which should not exceed a sufficiently small part (20% for example) of the human reaction time. For example, when an audio biofeedback application is planned to be used with athlete group and the athlete's reaction time is around 150 ms (Human Reaction Time 1970), it is reasonable to set the sampling frequency to 100 Hz or higher, so the feedback delay stays well below the athlete's reaction time. Gyroscope devices can send samples with much higher sampling rates, while the Qualisys™ optical tracking system real-time protocols has a limited rate of 60 Hz (Nilsson 2011).

Testing applications mostly run on modern laptops where signal processing power is sufficient. Signal processing time can be much shorter than sampling time and therefore negligible. Processing power can become an issue when a biofeedback application is planned to run on low-power wearable body-attached signal processing device or smartphone. In such cases the processing time may equal or even exceed the sensor sampling time. In the latter case the processing does not satisfy the requirements of the real-time operation any more.

Sensor samples are taken at equidistant times and form a synchronous sample data stream. Additional feedback loop delay can be accumulated by the improper setting of the asynchronous communication protocols between sensors, signal processing device and actuators.

When packetizing more than one signal sample in one data packets additional delay is generated automatically; such cases should therefore be omitted. For example, with the sensor sampling frequency of 100 Hz the sensor sampling time is 10 ms, and when packetizing 3 samples into one packet, the delay at the sensor is already 30 ms.

Packet loss during the transmission occurs when sensor data is streamed in real time over the unreliable communication channels, such as IEEE 802.11 and Bluetooth wireless links. For real-time streaming the communication protocols without retransmission should be used. For example, UDP transport protocol in the case of IP networks. Retransmissions are not welcome because the retransmitted sample would most probably arrive at the processing unit out of order and could interfere with the correct sensor signal analysis.

Communication protocols with asynchronous operation many times employ receiving buffers to smooth the synchronously streamed data, compensate the variable packet delays, and hence try to replay the data synchronously. This is achieved by introducing an additional playback delay at the receiver. The playback delay adds to the total feedback loop delay. If possible receiving buffers should be omitted or set to the lowest possible value.

Regarding the above, it is of outmost importance, that for each separate real-time biofeedback system, one chooses the adequate combination of equipment, sets parameters to the appropriate values, and chooses the communication protocols, that work favourably for the real-time operation of the biofeedback system.

6.6.2 Communication Demands of Sensors and Actuators

Sensors and actuators used in sport are very heterogeneous and can be grouped according to several criteria; measured quantity, delay demands, bit rates, etc. Table 6.7 lists the most used sensors and actuators with their corresponding bit rates or bit rate ranges and the delay constraints (Cavallari et al. 2014; Chen et al. 2011; Cao et al. 2009; Kos et al. 2016; Siddiqui et al. 2018; Umek and Kos 2016a)

One distinctive group of sensors measure low dynamic human physiological parameters such as temperature, heart rate, breathing, blood pressure, blood sugar,

Table 6.7 Bit rates and delay constraints for sensors and actuators used in sport and rehabilitation (Kos et al. 2018)

Sensor	Bit rate	Delay
Temperature sensor	<100 bit/s	Not critical
Heart rate sensor	<100 bit/s	Seconds
Oximeter	<100 bit/s	Seconds
CO_2 sensor	<100 bit/s	Seconds
Blood sugar sensor	<100 bit/s	Not critical
Blood pressure sensor	<100 bit/s	Not critical
ECG	20–100 kbit/s	<1 s
Accelerometer	1–200 kbit/s	<50 ms
Gyroscope	1–200 kbit/s	<50 ms
Magnetometer	1–200 kbit/s	<50 ms
Altimeter	<1 kbit/s	Not critical
Strain gage	1–50 kbit/s	<50 ms
MPEG4 camera	<10 Mbit/s	<50 ms
Actuator	Bit rate	Delay
Tactile actuator	<100 bit/s	<50 ms
Headphones/Loudspeaker (Voice)	50–100 kbit/s	<50 ms
Headphones/Loudspeaker (Audio)	<1 Mbit/s	<50 ms
Display (Video)	<10 Mbit/s	<50 ms

oxygenation, and others. Their common characteristic is low bit rate. Most of the above mentioned sensors measure the parameter with the frequency of up to 1 Hz and communicate it to the system even in longer time periods (5 s, 10 s, 60, etc.). The produced bit rate of physiological sensors is estimated to be below 100 bit/s.

With more dynamic physiological processes, like electrocardiogram (ECG), the entire signal at higher sampling frequencies is required. For example, ECG signal sampled at 1000 Hz with 24 bit precision, would produce a 24 kbit/s data flow. Even with more demanding physiological processes it can be estimated that the single sensor (device) does not produce a data flow of more than a few hundreds of kbit/s.

Sport sensors are most commonly associated with measuring the kinematic parameters of the performed activity. For those purpose accelerometers, gyroscopes, and magnetometers are commonly combined in one miniature inertial measurement unit (IMU). Different sensors in different devices operate at different sampling frequencies (from 10 Hz to 2000 Hz) and with different precision (from 12 to 24 bits). For example, a 9 DOF sensor device, operating in the above mentioned ranges, would produce data flows from 1.08 kbit/s to 432 kbit/s. When more than one IMU devices are used at the same time, the bit rate increases proportionally to the number of active IMU devices.

In addition to a well-known IMU devices, sport equipment can include integrated resistive and semiconductor strain gage sensors for indirect measurement of force, torque, pressure, and bend. Similarly to IMU, the bit rate depends on sampling frequency and precision. For example, a strain gage sensor with sampling frequency of 2000 Hz and 16 bit precision generates 32 kbit/s. Collective bit rate increases linearly with the number of active sensors.

The simplest actuators, such as buzzers (tactile) or beepers (audio), essentially use only one bit of feedback information, which tells the actuator to be active or idle. The bit rate of such simple actuators can be well below 100 bit/s. The situation is very different when the feedback signals are "natural" human sense signals (voice, audio, and video). Those signals are quite demanding and their bit rates can be anything between a few kbit/s (speech) to a few Mbit/s (video). Different modalities can be combined in the feedback information; for example, haptic and auditory (or video), where more than one haptic actuator can be used, but only one audio (or video) signal in the feedback.

By combining one or more physiological, one or more kinematic sensor, one or more sensor integrated into equipment, together with actuators, the required bit rates and delay constraints can be very high. The selection of the most appropriate wireless technology is of paramount importance for achieving a high enough quality of service of the biofeedback system operation.

6.6.3 Communication Technologies

Communication technologies enable signal and data transfer between the independent biofeedback system devices. The most appropriate and used for this task are

Table 6.8 Standardized wireless technologies with potential use in sport (Kos et al. 2018)

Technology	Frequency	Range	Bit rate	TX Power
Bluetooth 2.1 + EDR	2.4 GHz	10–100 m	1–3 Mbit/s	2.5–100 mW
Bluetooth 4.0 + LE	2.4 GHz	10 m	1 Mbit/s	2.5 mW
ZigBee	868 MHz and 2.4 GHz	10–100 m	20–250 kbit/s	1–100 mW
IEEE 802.11n	2.4 and 5 GHz	70 m	600 Mbit/s	100 mW
IEEE 802.11ac	5 GHz	35 m	6.93 Gbit/s	160 mW
IEEE 802.11ad	60 GHz	10 m	6.76 Gbit/s	10 mW
IEEE 802.11ah	900 MHz	1 km	40 Mbit/s	100 mW
IEEE 802.11af	54–790 MHz	>1 km	1.8–26.7 Mbit/s	100 mW
LoRaWAN	868–928 MHz	up to 100 km	250–5470 bit/s	1.5–100 mW

wireless communication technologies. An alternative to them are wired technologies, which are also commonly used in practice. In biofeedback systems, where implants are used, human body serves as the propagation channel (Cavallari et al. 2014). While wired communication channels usually do not present any serious limitations when used in biofeedback systems, and implants are more or less exotic at this time, we dedicate the most attention to wireless technologies.

Wireless technologies are, like sensors and actuators, very heterogeneous. Here we do not intend to give a detailed comparative analysis of all wireless technologies, but limit the discussion on the standardized and employed technologies that can be used for biofeedback applications in sport and rehabilitation.

Table 6.8 lists the most widespread wireless technologies for BAN, PAN, LAN and MAN (Metropolitan Area Network). Only the parameters relevant for further discussion are listed, details can be found in (Adame et al. 2014; Ayub et al. 2015; Baños-Gonzalez et al. 2016; Cao et al. 2009; Deslise 2015; Kwak et al. 2010; Lin et al. 2015; Movassaghi et al. 2014; Pyattaev et al. 2015; Zhang et al. 2014). The complete list of all wireless technologies and their variants that are available, under development or under testing, is much more extensive. Table 6.8 lists primarily the technologies that can be acquired on the market without major obstacles or the technologies that are expected to have commercial products available soon.

The selection of the most appropriate wireless technology depends heavily on the given architecture of the biofeedback system from Sect. 4.4 and the place of sensor signal processing from Sect. 6.6.1. The heterogeneity of wireless technologies and the variety of biofeedback system architectures suggests the use of multi-radio concepts (Hämäläinen et al. 2015).

Two main parameters are taken into consideration when selecting the most appropriate wireless technology for a biofeedback system in sport and rehabilitation: range (coverage) and bit rate. Both parameters are highly dependent on the selected biofeedback system architecture and place of sensor signal processing. Some of many examples of considerations when choosing the most appropriate wireless technology are:

- For sending the sensor data from an open space system architecture a LoRaWAN or a IEEE 802.11ah technologies are needed.
- ZigBee and LoRaWAN are not suitable for the transmission of high dynamic IMU signals.
- Bluetooth is suitable for personal biofeedback applications, but not for confined and open space biofeedback applications.

For biofeedback applications in sport the power and battery life are not of the primary concern. Contrary to sensor networks, batteries of sensor devices in sport are easily accessible for changing or recharging. Since the operation time of a biofeedback system in sport is limited by human endurance, battery life time can be much shorter than in some other sensor based applications, where batteries must last for days, months or even years. For this reason Table 6.8 lists not only the well-known sensor network technologies such as Bluetooth, ZigBee, and LoRaWAN, but also several LAN technologies with various bit rates and ranges. LAN technologies fill (a) the range gap between 100 m and 1 km, and (b) the bit rate gap above 2 Mbit/s.

LAN technologies are medium power, medium bit rate, and medium range. For example, a quick calculation shows that a small and lightweight battery with the 500 mAh capacity could support the operation of a LAN transceiver with a peak power of 100 mW for 5 to 10 h; more than enough for most professional sport use.

Not mentioned until now are the mobile wireless technologies such as GPRS, EDGE, 3G, and 4G. They are not widely employed in current sensor devices in sport. They are usable for terminal biofeedback systems, but due to their latency, they are not suitable for concurrent biofeedback systems. This may change with the promises of 5G standards (Wang et al. 2014). It is envisioned that 5G mobile networks will have much lower latencies, much more capacity, with much higher spectral efficiency, connecting the entire world by achieving communications between anybody, anything, anywhere, anytime, and with whatever device. But until this comes to be, we are bound to technologies which are available today.

Selection of Wireless Communication Technology

A large number of factors and constraints have to be considered when selecting the most appropriate elements of biofeedback systems in sport and rehabilitation. This is particularly true for the concurrent biofeedback systems that operate in real time.

For help with the selection process, we have composed Fig. 6.13 that illustrates the available bitrates of wireless technologies from Table 6.8 plotted against their ranges, bitrate ranges of various sensors and actuators from Table 6.7, and space constraints of feedback systems defined in the Introduction. Computational constraint is not addressed in detail here because it can be controlled to a high degree by the system designer, while the other two constraints are mostly the given properties of the feedback system. It should be noted that IEEE 802.11ah, IEEE 802.11af wireless technologies are trying to fill-in the gap for open space systems with kilometre ranges

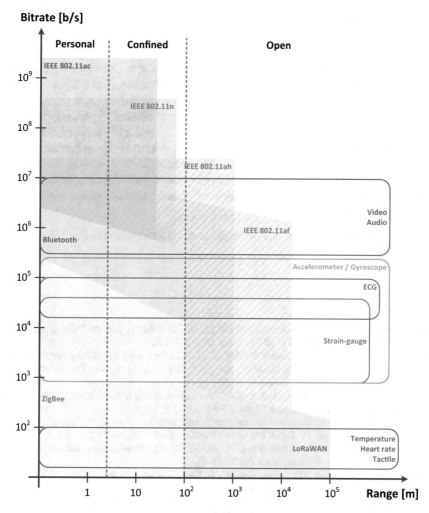

Fig. 6.13 Illustration of available bitrates of wireless technologies from Table 6.8 plotted against their ranges, bitrate ranges of various sensors and actuators from Table 6.7, and space constraints help with the selection of appropriate feedback system elements

and bitrates in Mbit/s, but at the time of writing they were not yet available in the market.

A number of important conclusions for the design of real-time feedback system can be drawn from Fig. 6.13., such as for example:

- Real-time feedback systems based on physiological parameters, such as temperature and heart rate, can be implemented in personal, confined, and open space by using LoRaWAN wireless technology.

- Usage of audio and video actuators in personal space systems is supported by Bluetooth, IEEE 802.11n, IEEE 802.11ac, IEEE 802.11ah, and IEEE 802.11af wireless technologies; in open space systems these actuators can be used to some extent by implementing IEEE 802.11ah, IEEE 802.11af wireless technologies; the problem is that the latter two technologies are standardized, but not yet available in the market.
- Inertial sensors (accelerometer and gyroscope) can be used in personal and confined space systems by using all listed wireless technologies, except LoRaWAN, but that is always true only for one such sensor; if there are more inertial sensors in the feedback system, some of the above technologies can prove as insufficient.
- Given the ZigBee based system in confined space, sensors for low dynamic physiological processes and a limited number of inertial and strain-gauge sensors with low sample rates can be used.

Many similar conclusions can be made based on information in Fig. 6.8.

References

Adame T, Bel A, Bellalta B, Barcelo J, Oliver M (2014) IEEE 802.11 AH: the WiFi approach for M2 M communications. IEEE Wirel Commun 21(6):144–152

Aggarwal P, Syed Z, Niu X, El-Sheimy N (2006) Cost-effective testing and calibration of low cost MEMS sensors for integrated positioning, navigation and mapping systems. In: Proceedings of XXIII FIG congress, Munich, Germany, vol 813

Aggarwal P, Syed Z, Niu X, El-Sheimy N (2008) A standard testing and calibration procedure for low cost MEMS inertial sensors and units. J Navig 61(2):323–336

Allan Variance (2003) http://www.allanstime.com/AllanVariance/. Accessed 30 June 2018

Allan DW (1966) Statistics of atomic frequency standards. Proc IEEE 54(2):221–230

Ayub K, Zagurskis V (2015) Technology implications of UWB on wireless sensor network-a detailed survey. Int J Commun Netw Inf Secur (IJCNIS) 7(3)

Baños-Gonzalez V, Afaqui MS, Lopez-Aguilera E, Garcia-Villegas E (2016) IEEE 802.11 ah: a technology to face the IoT challenge. Sensors 16(11), 1960

Cao H, Leung V, Chow C, Chan H (2009) Enabling technologies for wireless body area networks: a survey and outlook. IEEE Commun Mag 47(12)

Cavallari R, Martelli F, Rosini R, Buratti C, Verdone R (2014) A survey on wireless body area networks: Technologies and design challenges. IEEE Commun Surv Tutor 16(3):1635–1657

Chen M, Gonzalez S, Vasilakos A, Cao H, Leung VC (2011) Body area networks: a survey. Mob Netw Appl 16(2):171–193

Deslise JJ (2015) What's the difference between IEEE 802.11 af and 802.11 ah? Microw RF 54:69–72

Dixon-Warren SJ (2010) Motion sensing in the iPhone 4: MEMS Accelerometer. MEMS Journal

Dixon-Warren SJ (2011) Motion sensing in the iPhone 4: MEMS gyroscope. MEMS Journal

El-Diasty M, Pagiatakis S (2008) Calibration and stochastic modelling of inertial navigation sensor errors. J Glob Position Syst 7(2):170–182

El-Sheimy N, Hou H, Niu X (2008) Analysis and modeling of inertial sensors using Allan variance. IEEE Trans Instrum Meas 57(1):140–149

Grewal M, Andrews A (2010) How good is your gyro [ask the experts]. IEEE Control Syst 30(1):12–86

Hämäläinen M, Paso T, Mucchi L, Girod-Genet M, Farserotu J, Tanaka H, ... Ismail LN (2015) ETSI TC SmartBAN: overview of the wireless body area network standard. In: 2015 9th international symposium on medical information and communication technology (ISMICT), pp 1–5. IEEE

Hongwei S, Yuli L, Guangfeng C (2010) Relations between the Standard variance and the Allan variance. In: 2010 international conference on computational and information sciences, pp 66–67. IEEE

Human Reaction Time (1970–1979) The Great Soviet Encyclopedia, 3rd edn

IEEE (1999) IEEE standard specification format guide and test procedure for linear, single-axis, non-gyroscopic accelerometers

Jiang C, Xue L, Chang H, Yuan G, Yuan W (2012) Signal processing of MEMS gyroscope arrays to improve accuracy using a 1st order markov for rate signal modeling. Sensors 12(2):1720–1737

Kos A, Milutinović V, Umek A (2018) Challenges in wireless communication for connected sensors and wearable devices used in sport biofeedback applications. In: Future generation computer systems

Kos A, Tomažič S, Umek A (2016a) Suitability of smartphone inertial sensors for real-time biofeedback applications. Sensors 16(3):301

Kos A, Tomažič S, Umek A (2016b) Evaluation of smartphone inertial sensor performance for cross-platform mobile applications. Sensors 16(4):477

Kwak KS, Ullah S, Ullah N (2010). An overview of IEEE 802.15. 6 standard. In: 2010 3rd international symposium on applied sciences in biomedical and communication technologies (ISABEL), pp 1–6. IEEE

Land DV, Levick AP, Hand JW (2007) The use of the Allan deviation for the measurement of the noise and drift performance of microwave radiometers. Meas Sci Technol 18(7):1917

Leland RP (2005) Mechanical-thermal noise in MEMS gyroscopes. IEEE Sens J 5(3):493–500

Lin JR, Talty T, Tonguz OK (2015) On the potential of bluetooth low energy technology for vehicular applications. IEEE Commun Mag 53(1):267–275

Liu M (2013) A study of mobile sensing using smartphones. Int J Distrib Sens Netw 9(3):272916

Looney M (2010) A simple calibration for MEMS gyroscopes. EDN (Electri Des News) 55(9):21

Mohd-Yasin F, Korman CE, Nagel DJ (2003) Measurement of noise characteristics of MEMS accelerometers. Solid-State Electron 47(2):357–360

Movassaghi S, Abolhasan M, Lipman J, Smith D, Jamalipour A (2014) Wireless body area networks: A survey. IEEE Commun Surv Tutor 16(3):1658–1686

Nilsson L (2011) QTM Real-time Server Protocol Documentation Version 1.9. http://qualisys.github ub.io/rt-protocol/. Accessed 10 Sept 2015

Prikhodko IP, Trusov AA, Shkel AM (2013) Compensation of drifts in high-Q MEMS gyroscopes using temperature self-sensing. Sens Actuators, A 201:517–524

Pyattaev A, Johnsson K, Andreev S, Koucheryavy Y (2015) Communication challenges in high-density deployments of wearable wireless devices. IEEE Wirel Commun 22(1):12–18

Qualisys, Motion Capture System (2018). http://www.qualisys.com. Accessed 20 June 2018

Shaeffer DK (2013) MEMS inertial sensors: A tutorial overview. IEEE Commun Mag 51(4):100–109

Siddiqui SA, Zhang Y, Lloret J, Song H, Obradovic Z (2018) Pain-free blood glucose monitoring using wearable sensors: recent advancements and future prospects. In: IEEE reviews in biomedical engineering

ST Microelectronics (2009) M.E.M.S. digital output motion sensor ultra low-power high performance 3-Axes "Nano" Accelerometer, LIS331DLH Specifications

ST Microelectronics (2010) MEMS motion sensor: ultra-stable three-axis digital output gyroscope. L3G4200D Specifications, ST Microelectronics, Geneva, Switzerland

ST Microelectronics (2011) Everything about STMicroelectronics'3-Axis Digital MEMS Gyroscopes, TA0343, Technical article. ST Microelectronics. July, 36

Stančin S, Tomažič S (2014) Time-and computation-efficient calibration of MEMS 3D accelerometers and gyroscopes. Sensors 14(8):14885–14915

Stockwell W (2004) Bias stability measurement: Allan variance. http://www.moog-cross-bow.co m/Literature/Application_Notes_Papers/Gyro_Bias_Stability_Measurement_using_Allan_Vari ance.pdf. Accessed 26 Dec 2015

Umek A, Kos A (2016a). The role of high performance computing and communication for real-time biofeedback in sport. Math Probl Eng 2016

Umek A, Kos A (2016b) Validation of smartphone gyroscopes for mobile biofeedback applications. Pers Ubiquit Comput 20(5):657–666

Wang CX, Haider F, Gao X, You XH, Yang Y, Yuan D, … Hepsaydir E (2014) Cellular architecture and key technologies for 5G wireless communication networks. IEEE Commun Mag 52(2), 122–130

Weinberg H (2011) Gyro mechanical performance: the most important parameter. Technical Article MS-2158

Woodman OJ (2007) An introduction to inertial navigation (No. UCAM-CL-TR-696). University of Cambridge, Computer Laboratory

Zhang Y, Sun L, Song H, Cao X (2014) Ubiquitous WSN for healthcare: recent advances and future prospects. IEEE Internet Things J 1(4):311–318

Chapter 7
Applications

7.1 Application Variety

Given a large number of different sports and rehabilitation therapies, a great variety of different biofeedback applications is needed. As discussed in Chap. 4, biofeedback system elements can be put together in many different combinations according to the properties, requirements, usage, constraints, and other factors important to the application. The most influential factors are the type of feedback (terminal, cyclic, and concurrent) and the intended functionality of the biofeedback application (user, instructor, and cloud). For example, the structure of the biofeedback application (compact or distributed) depends on its intended functionality and its expected physical extent.

The rapid development of technology has made feasible many biofeedback system implementations that were not possible some time ago. The technologies used in biofeedback systems are presented in Chap. 6 where some of their limitations are also discussed. The task of the biofeedback application developer is to choose the most appropriate biofeedback system architecture implement it with the most optimal choice of technologies for each of the system elements. General properties of sports and rehabilitation are discussed next.

7.1.1 Properties and Requirements

Type of sport, a particular sport discipline or the type of rehabilitation therapy are the major deciding factors for the correct choice of the most appropriate technologies used in development of the biofeedback applications. As already said, the number of sport disciplines and rehabilitation therapies is far too large to be able to discuss all the different biofeedback system architectures and technological challenges that a developer is facing when designing a biofeedback application.

© Springer Nature Switzerland AG 2018
A. Kos and A. Umek, *Biomechanical Biofeedback Systems and Applications*,
Human–Computer Interaction Series, https://doi.org/10.1007/978-3-319-91349-0_7

Similarly to the classification of biofeedback system architectures, presented in Sect. 4.5, sport disciplines and rehabilitation therapies can be classified from the technical aspect according to the following criteria: (a) place (fixed, bounded, unbounded), (b) number of users (individual, group, mass), (c) movement dynamics (low, medium, high), (d) movement type (defined single aperiodic movement, cyclic-periodic movement, not defined free movements), (e) used equipment (none, simple, complex), (f) environment (indoor, outdoor, water, etc.), and (g) analysis complexity (low, medium, and high).

To shed some light on the biofeedback application design, we present some typical biofeedback application scenarios that are based on the above classification of sport disciplines and rehabilitation therapies.

7.1.2 Typical Application Scenarios

Typical application scenarios base on three of the classification criteria of sport disciplines and rehabilitation therapies; movement dynamics, number of users, and analysis complexity. Three application scenarios are presented: low dynamic scenario, high dynamic scenario, and high dynamic multiple sensor and multiple user scenario.

Low-Dynamic

Low dynamic biofeedback applications are using signals from sensors with low sampling frequencies. Such applications are not demanding in terms of communication; bit rates, even with a larger number of sensors, reach up to several kbit/s.

Such applications can be implemented in any system architecture discussed in Chap. 4 and by using several different technologies. For example, user architecture is implemented with Bluetooth LE, confined space architecture with IEEE 802.11af, and open space architecture with LoRaWAN; see Table 6.8 for reference on wireless technologies.

Processing power in low dynamic application is generally not a problem because low bit rate streams do not produce large amounts of data for processing and analysis.

An example of usage in sport is a biofeedback application that gives feedback based on measuring and analysis of physiological parameters of a user during running on a treadmill in fitness, on a track on stadium, or in the nature. An example of usage in rehabilitation is a biofeedback application for correct gait restoration after injury.

High-Dynamic

High dynamic biofeedback applications are using signals from sensors with medium to high sampling frequencies. Such applications are more demanding. Their bit rates can easily reach several hundred kbit/s; when more sensors are used simultaneously, even several Mbit/s.

Such applications can be more easily implemented in personal or confined system architectures discussed in Chap. 4 where many wireless technologies satisfy their communication demands (see Table 6.8). For the open space architecture a relay over the gateway or mobile device with mobile network connectivity is possible, providing a high enough bit rate is available.

Processing power in high dynamic application can be a problem, especially if concurrent feedback functionality is required. The problem arises particularly with wearable end embedded devices with low processing power. When using a computer or even more powerful processing device, the processing is generally not a problem.

An example of high dynamic biofeedback application in rehabilitation is giving visual, graphically animated feedback to the person rehabilitating its gait. An example in sport is giving concurrent auditory feedback to a gymnast performing practice on a vaulting horse.

Multiple Sensors and Multiple Users

A biofeedback application using signals from a combination of sensors with high sampling rates, sport equipment sensors, and actuators with video based feedback can be very demanding. Bit rates can exceed 10 Mbit/s when a number of sensors are on the user's body, a number of sensors integrated into the sport equipment, and 3D graphical video is used for the feedback. Even the personal and confined space architectures can be implemented only through a limited number of wireless technologies (see Table 6.8). Open space architectures are not viable.

Processing power in such applications can be a problem. A powerful computer or even a supercomputer is needed, if concurrent feedback functionality is required. Processing on wearable devices is not viable for such scenarios.

An example of usage in sport is giving concurrent video feedback to a skier with a personal architecture of the system (all devices on the skier's body or/and in the backpack).

Even more demanding scenarios are possible when using such biofeedback applications in group sports. For example, the scenario for a high performance real-time biofeedback application for a football match includes 22 active players, 3 judges, 10 sensors per person, 1000 Hz sampling rate, 9 DoF sensor data, each with 16 bits per measured quantity. The nett bit rate of all sensors achieves 36 Mbit/s in the direction to the processing unit. Implementing a real-time biofeedback application in this scenario is an extremely demanding task for all elements of the system; from sensors and processing unit, to communication channels.

7.2 Application Examples

A large number of biofeedback applications from the domain of sport and rehabilitation have been developed to date. The pace of new applications releases is growing from year to year. The main reason for that is the development of new, better technical equipment used in such systems. Also, such equipment is becoming cheaper and more affordable every year.

Table 7.1 List of presented applications and their main properties according to the classification given in Sect. 4.5 and Fig. 4.4

Application	Functionality	Extent	Structure	Feedback	Specifics
Golf swing trainer	User	Personal confined	Compact distributed	Concurrent	Gesture user interface
Smart golf club	Instructor	Confined	Distributed	Concurrent terminal	Shaft bend measurement
Smart ski	User instructor	Personal open	Compact	Concurrent cyclic terminal	Complex with many sensors
Water sports	User instructor	Open	Distributed	Cyclic terminal	Water environment
Swimming rehabilitation	User Instructor	Personal confined	Compact distributed	Concurrent cyclic	Healthcare

Biofeedback applications presented in this chapter are listed in Table 7.1. The table includes several applications in different sport disciplines and one application in rehabilitation therapy.

7.3 Golf Swing Trainer Application

The *Golf swing trainer* application is an example of a wearable training application with concurrent biofeedback and a gesture user interface. The application is aimed at helping golfers to correct unwanted head movements in real time during the execution of their golf swings.

7.3.1 Objective and Functionality

The application is designed and implemented to prove the concept of accelerated motor learning with the help of augmented concurrent biomechanical biofeedback. The application is using the example of a golf swing movement, which is a static activity with well-defined goals. The main functionality of the application is real-time golf swing analysis and feedback, with a focus on the golfer's head movements.

The application provides user functionality and can be implemented in a compact or in a distributed biofeedback structure (see Sect. 4.5). A smartphone enables the simple implementation of the compact structure, while the distributed structure can be easily implemented using a laptop. Compact structure corresponds to the personal and distributed structure corresponds to the confined space system. Current high-end smartphones include all of the required biofeedback system components; sensors, processing device, and actuators. For more demanding uses in terms of the number of sensors, wide dynamic range, device size, or other constraints, dedicated devices are required.

The operation of the application is driven by user gestures. Gestures are defined as a user's body movements, which are detected via their characteristic motion sensor responses. The details of the gesture user interface are discussed in the following sections.

7.3.2 System Architecture and Setup

The application consists of a wearable processing device, one or more body-attached motion sensors with 3-axis accelerometers and 3-axis gyroscopes, and an audio device (headphones) for biofeedback. In the case of the compact structure smartphone performs the functions of sensing and processing and in connection to the headphones also the feedback function. In the case of the distributed structure the function of processing is transferred to the laptop, which is wirelessly connected to smartphone and headphones. The distributed structure of the system is presented in Fig. 7.1.

Basic Concepts

The operation of the application is based on the segmentation of the complex golf swing movement pattern into a series of simple, short-time movement patterns. It was shown in Stančin and Tomažič (2013) that the early detection of improper motion in golf swings is possible via a post-swing analysis of recorded motion sensor readings.

Our real-time biofeedback concept is based on the hypothesis that incorrect movement patterns in a golf swing often lead to distinct unwanted head movements. Consequently, head movements are often an indicator of incorrect golf swing execution. This claim is also supported by observations of the world's best golf players (Doyle

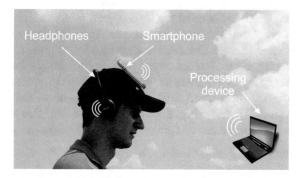

Fig. 7.1 Distributed structure of the biomechanical biofeedback system: headphones provide audio feedback to the user, smartphone with an integrated 3-axis accelerometer and 3-axis gyroscope provides sensing, and laptop serves as a processing device that is wirelessly connected to the smartphone and headphones

2015; Woods 2009) the majority of whom keep their heads practically still until just after the moment when the head of the club hits the ball.

With the appropriate attachment of the motion sensor to the cap of the golf player, as seen in Fig. 7.1, good detection repeatability can be achieved for various 3D head movements measured from the static start position during the swing setup phase.

Gesture User Interface

A gesture-driven user interface eases application usage and is able to detect different swing phases. In addition, the application provides users with real-time audio feedback during the execution of the swing. First, the application guides the user in achieving the correct swing setup position. Second, the application signals user head movement errors in real time during the swing. After the swing, the application lists possible head movement errors and draws various detailed head movement diagrams on the laptop or smartphone screen.

The user operates the application using an application-specific gesture user interface (GeUI). To detect gestures, the GeUI uses the same sensor signals as does the analytical component of the application. Gestures are defined as specific, previously agreed-upon head movements. For example, the application implements gestures for triggering the beginning of a new swing, saving swing data, and deleting swing data. A gesture user interface is mandatory, if the application, which runs on a wearable device, is to be easy to use and non-distracting to the user.

The use of certain of the abovementioned GeUI gestures during the process of a golf swing setup is briefly illustrated in Fig. 7.2. The reference in this process is the vertical alignment of the smartphone, which indicates the correct final swing setup position. In Fig. 7.2a, the head is in the upright position, where the angle between the smartphone and the vertical reference is typically between 40 and 50°; thus, the application is in the standby state. In Fig. 7.2b, the user is moving from the upright to the setup position, and the application detects this characteristic movement as a gesture that triggers the beginning of the setup phase.

During the setup phase, the application guides the user towards the proper setup position by changing the volume of the audio signal (noise) sent to the user. The noise volume is proportional to the user's distance from the proper setup position. The application allows for positions that deviate by a few degrees from the vertical alignment of the smartphone. For amateur players, the threshold values are typically set at approximately 5°; for professional players, this value is typically 2–3°. When the correct setup position is reached, as in Fig. 7.2c, the absence of head movements is detected, and the user is asked to begin an upswing. The application is then prepared for the swing, and the user receives a characteristic audio message.

Real-time Signal Analysis and Concurrent Biofeedback

The analysis of sensor data and the generation of audio biofeedback signals are performed in real time throughout the entire duration of the golf swing, as presented in Fig. 7.4. All sensor data and analysis results are recorded for later post-processing, which is used primarily for the development of and improvements to the application as well as to evaluate the effects of biofeedback.

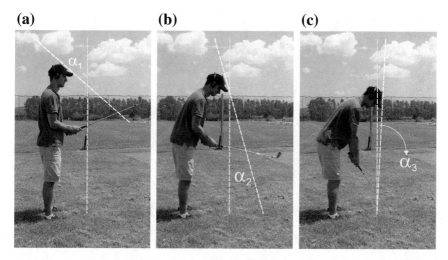

Fig. 7.2 Process of a golf swing setup: **a** upright position, **b** moving from the upright to the setup position, and **c** setup position. The white dotted lines indicate smartphone orientations

Fig. 7.3 Main golf swing phases: **a** setup position, **b** backswing, **c** top of the backswing, **d** impact, and **e** follow-through. The white dotted lines indicate the smartphone orientation

Figure 7.3 shows the main golf swing phases, during which the application analyses head movements and detects possible errors. The swing begins when the proper setup position is reached; see Figs. 7.2c and 7.3a. The user proceeds through the consecutive swing phases from the backswing (Fig. 7.3b) to the follow-through (Fig. 7.3e), during which the application issues the user audio feedback when errors are detected. Figure 7.3c shows an example of an excessive head rotation that would trigger the error audio feedback signal.

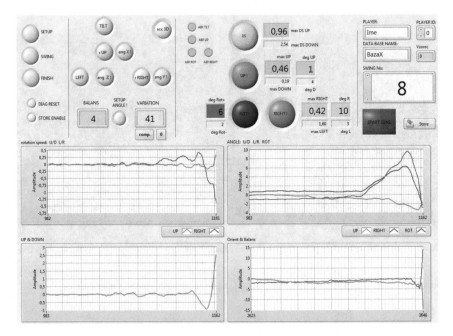

Fig. 7.4 The application window offers additional information on the swing by means of numerous error indicators, head movement signal peaks, and head movement diagrams (Umek et al. 2015)

The application is not designed to identify the exact cause of the incorrect movement; rather, it merely alerts the user to golf swing errors concurrently with the swing execution. When players are aware of the cause of their swing errors, they can easily understand the feedback signals and consequently act on them.

During golf swing training, the user should focus primarily on the acoustic feedback signal. Additional information on the swing, which is simultaneously shown in numerous diagrams by means of error indicators and peak value detectors, as shown in Fig. 7.4, is primarily intended for the instructor. The user can easily review this information on a laptop after the swing or at any later time.

Application Evaluation

The application was first evaluated in terms of correct and precise application operation (the first and second testing stages) and then in terms of the usability of the real-time biofeedback for the correction of errors during the performance of a golf swing (the third testing stage).

The precisions of the accelerometer and gyroscope are primarily affected by their biases, which induce errors in their derived spatial and angular positions (see Chap. 6 for details). The measured bias values of the smartphone accelerometer are in the range of ± 12 mg$_0$, and the measured bias values of the gyroscope are in the range of ± 1.15 deg/s (Kos et al. 2016). The accelerometer data are used only for determining the standstill head position, and the derived angular error is less than $0.7°$. In the worst case, the derived angular error of the gyroscope does not exceed $3°$ during the execution of a golf swing.

The precisions of the accelerometer and the gyroscope are considerably improved after bias compensation (Kos et al. 2016). Gyroscope bias compensation can be performed with the smartphone in a standstill position. Shortly after compensation, the calculated variations in the mean bias values are 0.25 mg$_0$ for the accelerometer and 25 mdeg/s for the gyroscope. The bias drift over one hour increases both angular positioning errors. The vertical positioning angle error is less than 0.25, and the predicted gyroscope rotation angle error is less than 0.35 deg/s. Thus, in the worst case, the derived gyroscope angular error does not exceed $1°$ during the execution of a golf swing.

The sensors are more accurate if their biases are compensated, but even uncompensated sensor readings are acceptable for the proposed application, as the signal analysis during a typical golf swing requires only two to three seconds.

7.3.3 Results

The application was tested in several player groups with different levels of golf skills. The first testing stage involved recording a number of consecutive swings performed within a time frame of 10 min by a professional golf player with highly consistent swing execution in terms of movement repeatability. The gyroscope bias variations during a time period of 10 min are negligible; thus, we can consider that the bias remained constant during the test and that the induced angular error is the same for all test plots shown in Fig. 7.5. The maximal induced gyroscope angular error during swing execution is $1°$, which is one order of magnitude lower than the biofeedback alarm threshold values. From the results presented in Fig. 7.5, it can be concluded that the application functions correctly and that the precision of the sensor signals is sufficiently high for analysis of the required quality. The records acquired for the professional golf player can also serve as the source and model in the setting of the threshold values for triggering and the state signals used by the application.

In the second stage, amateur golf players with different incorrect movement patterns were tested, as reflected in the data indicating unwanted excessive head movements. Figure 7.6 shows the head rotation signals for golf swings of an experienced amateur player. As expected, the swing consistency is much lower than that of the professional golf player.

In the third testing stage, the influence of biofeedback was tested on amateur golf players who all exhibited excessive head movements during their golf swings. Head movement errors were communicated to the users in real time through the audio channel in the form of a binary acoustic signal.

Figure 7.7 displays one example of the positive influence of biofeedback on one of the tested amateur players. It shows a comparison of the average swing execution before and after the application of biofeedback. The player first performed a series of swings without biofeedback, and the average result is plotted in Fig. 7.7 as a dashed line. Next, the player performed another series of swings with biofeedback, and the average result is plotted in Fig. 7.7 as a solid line.

The first subjective application evaluations based on the opinions of a group of fourteen amateur players indicate that the provided audio biofeedback aided in correcting their unwanted excessive head movements during their golf swings. The learning process for most of these players was more rapid than anticipated.

7.3.4 Discussion

This application has demonstrated that biofeedback applications are feasible using smartphones and their integrated motion sensors. It has also been confirmed that the sensor data obtained from such sensors are sufficiently accurate for short-time movement analysis. It is demonstrated that the developed hands-free gesture user interface is an indispensable element of the application because it alleviates the demands of user interaction with the application. Field test results indicate that the *Golf Swing Trainer* is an efficient tool for the correction of head movement errors by amateur golf players who experience difficulty with swing consistency.

Most importantly, the results of tests conducted on small groups of users indicate that concurrent biofeedback can accelerate motor learning process.

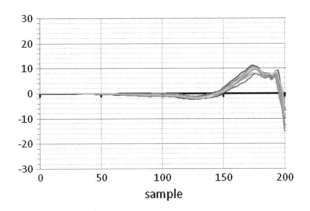

Fig. 7.5 Head rotation angle [degrees] in the *Right-Left* direction for the last 200 swing samples from a professional golf player. The head movement is consistent for all executed swings (Umek et al. 2015)

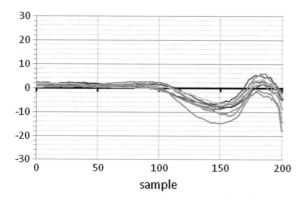

Fig. 7.6 Head rotation angle [degrees] in the *Right-Left* direction for the last 200 swing samples from an amateur golf player. The head movement is inconsistent (Umek et al. 2015)

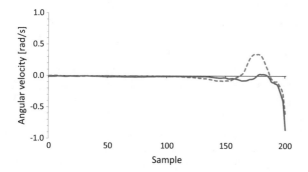

Fig. 7.7 Benefit of biofeedback for an amateur golf player. The graph shows a comparison of a player's head rotation speed [rad/s] in the *Right-Left* direction without biofeedback (dashed line) and with biofeedback (solid line). The curves represent the averages calculated from twenty swings performed without biofeedback and twenty swings performed with biofeedback (Umek et al. 2015)

7.4 Smart Golf Club Application

The *Smart golf club* application is an example of a biofeedback system with concurrent or terminal biofeedback based on smart sport equipment. The application is aimed at helping golfers to improve their golf swing execution. It alerts its users about improper swing parameters during the swing execution.

7.4.1 Objectives and Functionality

The application is designed for amateur and professional golf players and provides the instructor functionality (see Sects. 4.4 and 4.5). Amateurs can benefit primarily from its concurrent feedback that enables them to accelerate motor learning of the proper golf swing. Professionals can benefit primarily from application's terminal feedback that enables them to monitor their swing execution through time or to become aware of some minor changes or possible errors in their swing execution. The application is implemented as a distributed structure system in confined space.

7.4.2 Background

The main focus of research and development in sport is dedicated to wearable devices used for many different applications and purposes; from simple state detection devices to highly complex expert systems in professional sport (Lightman 2016; Yu et al. 2016).

In many sports the equipment is an inseparable element of the action. Athletes use sport equipment as a tool or medium through which their energy and actions are transferred into a desired result. Measuring and quantifying the actions of the athlete and the response of the sport equipment is expected to prove beneficial to athlete performance improvement. For the detailed quantification of sport activities a number of physical and physiological quantities should be measured simultaneously; from heart rate and body temperature, to exerted forces, material bends, accelerations, and rotation speeds, among others.

At this time wearable sport devices and equipment predominantly include motion sensors: accelerometers, gyroscopes and sometimes also magnetometers (Chambers et al. 2015; Hsu et al. 2016; Mendes et al. 2016; Yu et al. 2016). Examples of sport equipment with integrated IMU, such as smart tennis racket, smart electrical baseball bat, smart golf club, smart ball, are already on the market or are expected to be available in a few years (Lightman 2016). While motion sensors are sufficient to measure movement dynamics and short term positions, but they cannot reliably measure other important quantities needed in sports, such as force and bend (Abdul Razak et al. 2012; Mendes et al. 2016). Important complementary sensors in sport are integrated into sport equipment. The use of the concepts of sensor fusion and smart sensors can be applied to various sports and types of analyses (Mendes et al. 2016).

The use of inertial sensors is not new to golf. There are many applications, such as Ahmadi et al. 2014, Mitsui et al. 2015, Najafi et al. 2015, Nam et al. 2014, Naruo et al. 2013, Stančin and Tomažič 2013, Ueda et al. 2013, that use accelerometers, gyroscopes, or both, for golf swing tracking any analysis. One example of the use of optical tracking system for golf is presented in Jakus et al. (2017) where authors use a professional optical system Qualisys for detection of technical errors in putting practice. Such optical systems are unfortunately very expensive and highly impractical for the field use. Our primary focus was with strain gage sensors, which are not commonly used for this purpose.

The most often use of IMU devices in golf is body and/or club motion tracking. Ahmadi et al (2014), Nam et al. (2014), and Ueda et al. (2013) use accelerometers and gyroscopes attached at the grip end of the club for its position and orientation tracking. Naruo et al. (2013) use IMU device attached to the player's hand for the advice on the selection of the most appropriate club. One example of IMU device attached to the putter head is presented in Jansen et al. (2015). The most limiting factor in the above papers is the dynamic range of accelerometers and gyroscopes in regard to the suitable place of attachment: (a) if sensors are attached close to the club head, the motion dynamics (except with putter) will exceed their measuring range; (b) if sensors are attached close to the club grip they cannot measure the club shaft bend. Our system complements IMU device with strain gage sensors. It is capable of measuring the club shaft bend with strain gage sensors and the motion of the club with IMU device. The primary goal of our research is not golf swing tracking, but detection of any deviation between the reference swing and the swing under analysis.

Strain gage sensors have been used in golf as presented in Betzler et al. (2012), Choi et al. (2016), and Shyr et al. (2014). All studies are oriented primarily to the understanding of golf club performance and not to the analysis of the golf swing execution as it is the case in our application.

Our application uses sensors complementary to IMUs, which can be integrated into sport equipment. We present the general concepts, properties, and effects connected to the use of complementary sensor. We present this topic in the form of feasibility study of different sensors used for a golf swing analysis. For example, to accurately record the bend of the club shaft at impact, which is one of the important parameters of the golf swing, IMU is not appropriate because: (a) if mounted near the club grip, it is not accurate enough, (b) if mounted near the club head, the acceleration exceed its measuring range and the mechanical stress is too high. An alternative type of sensor should be used to measure the golf club shaft bend (Betzeler et al. 2012). Strain gage sensors have been found appropriate for the job as they measure the strain in the material. In our case the strain is caused by club shaft bending during the swing. Strain gage sensors are small-size and lightweight and fulfil the requirement for the sport equipment sensor.

7.4.3 System Architecture and Setup

The selection of sensors for the smart golf club is bounded by the following conditions. The golf swing is a high dynamic movement where the accelerations of the golf club parts can reach very high values far exceeding the available accelerometer measurement ranges that are typically ± 16 g_0. Similar observations are valid for the gyroscope as well. When the club head hits the ball, or worse, when the club hits the ground, the impact causes high frequency vibrations. Such vibrations can be measured only with sensor devices allowing high sampling frequencies. Bending is very important in golf swing; improper bending indicates improper swing. For this purpose we chose to use the strain gage sensors that measure golf club shaft strain that is directly dependent on the shaft's bending. The strain is a relative change of the dimensions of the material under load. It is defined as Dx/x, where x is a dimension in [m]. The unit for strain is dimensionless and is commonly presented as e. We have chosen to put the accelerometer and gyroscope sensor device just below the club grip and two orthogonally placed strain gage sensors in the upper section of the golf club shaft.

System Setup

The smart golf club measuring system presented in this paper includes: (a) two single grid strain gage sensors SGD-3/350-LY11 from Omega, which measure the golf club shaft bend, (b) 3-axis MEMS accelerometer and 3-axis MEMS gyroscope, which measure acceleration and angular speed of the golf club. The latter two sensors are a part of the independent Shimmer 3 IMU equipped with Bluetooth communication interface. Strain gage sensor signals are acquired by the National Instruments cRIO professional measurement system with module 9237. The measurement results gained by the strain gage, accelerometer, and gyroscope sensors are monitored and validated by the professional high-precision optical tracking system Qualisys (2018).

The smart golf club application was tested in several phases: sensor mounting and performance testing, sensor calibration, sensor validation, static tests, field tests. All experiments, except the last one, were conducted in the laboratory environment. The experimental system elements are described in the following subsections.

Sensors

For measuring the acceleration and rotation of the golf club we are using Shimmer 3 IMU with integrated 3-axis accelerometer and 3-axis gyroscope. According to our outdoor field test experience, Shimmer 3 device can wirelessly stream sensor data up to sampling frequencies of 512 Hz, which was used in our experiments. The accelerometer's dynamic range is up to ± 16 g_0 and the gyroscopes dynamic range is up to ± 2000 deg/s. The precision of both is 16 bits per sample. In the experiments the Shimmer 3 device is fixed to the club's shaft just below the grip as seen in Fig. 7.8.

Measuring of the golf club bend is performed by two strain gage sensors that are pasted parallel to the shaft's axis and orthogonally to each other as seen in Fig. 7.9. They are placed a few centimetres below the Shimmer 3 IMU. The *front* strain gage sensor measures the bend in the direction from and to the golf player (toe up–toe down); the *side* strain gage sensor measures the bend in the left-right (lead-lag) direction.

Optical Reference System

We used the optical motion capture system Qualisys™ (Qualisys Inc.) as a reference for the 3D rigid body tracking. Qualisys™ is a professional, high-accuracy tracking system Qualisys (2018) with eight Oqus 3 + high-speed cameras that offers real-time tracking of multiple predefined rigid bodies. The placement of the two independent rigid bodies, each with three reflective markers, can be seen in Fig. 7.8. The position of both rigid bodies in 3D space is shown in Fig. 7.10. The global coordinate system of the test space and two independent local coordinate systems of rigid bodies can also be seen in Fig. 7.10.

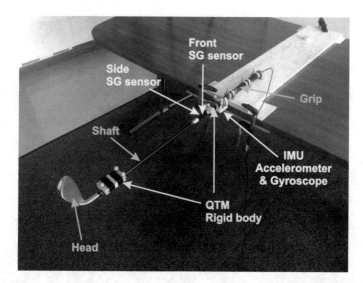

Fig. 7.8 Smart golf club equipped with various sensors fixed to the shaft at different positions: **a** the IMU device with 3-axis accelerometer and 3-axis gyroscope is just below the grip, **b** two SG sensors pasted orthogonally to each other along the shaft's axis, **c** Two independent rigid bodies used by the QTM optical system are fixed at the top and at the bottom of the shaft (Umek et al. 2017)

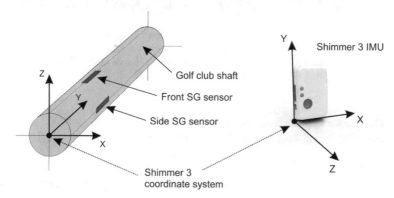

Fig. 7.9 Graphical representation of the smart golf club shaft with two orthogonally placed strain gage sensors that are measuring the bends of the shaft in two orthogonal directions. The front SG senor is placed in the direction of the Z axis and side SG sensor in the direction of the X axis of the Shimmer 3 IMU device; both measured from the axis of the shaft (Umek et al. 2017)

The exact position and orientation of the rigid body is captured by the Qualisys Track Manager (QTM) software application that defines the global coordinate system, determines the absolute position of each tracked marker, and calculates the orientation of rigid bodies. Relative rotations between bodies can be calculated as

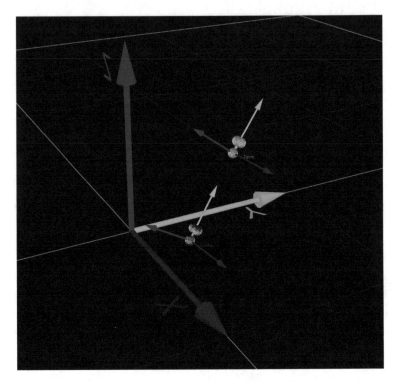

Fig. 7.10 3D rigid body tracking with Qualisys Track Manager. The positions of two independent rigid bodies attached to the golf club shaft at one point in time are shown in the global coordinate system XYZ. Each rigid body is composed of three markers and has an independent local coordinate system with the origin in one of the markers (Umek et al. 2017)

well. The marker capture frequency of the QTM can be as high as 1000 Hz, its real-time streaming operation, needed in our system, is limited to 60 Hz (Qualisys 2018).

Sensor Signal Acquisition and Processing

Sensor signals are synchronized and processed by the distributed LabVIEW™ application running simultaneously on the laptop and cRIO platform. Laptop and cRIO are interconnected by Wi-Fi connection. The sampling frequency of the main LabVIEW module running on the cRIO is 500 Hz. The application receives three independent data streams:

- QTM application streams the real-time rigid body orientation over the Ethernet connection to the LabVIEW module running on the laptop with the frequency of 60 Hz.
- Shimmer 3 IMU streams the real-time accelerometer and gyroscope sensor signals over the Bluetooth connection to the LabVIEW module running on the laptop with the frequency of 512 Hz.

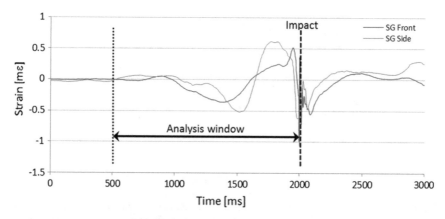

Fig. 7.11 Recording and analysis window of a single swing. Swing recording time is 3 s with the impact point set at 2 s. The analysis window has the width of 1.5 s (Umek et al. 2017)

- Strain gage sensors are connected by wire to the strain gage module NI 9237 inserted into the cRIO platform. The sampling frequency for strain gage signals is 500 Hz.

After real-time streamed sensor signals are aligned by their impact samples, they are segmented into separate swing signals. Each swing signal contains 1500 samples, where impact sample is always at index 1000. At the main application sampling frequency of 500 Hz the duration of each recorded swing is 3 s and impact point is at 2 s. The relations between the recording and analysis window are presented in Fig. 7.11.

In the graphs presented later in this application only swing signal samples between indexes 250 and 1000 (impact) are plotted. This corresponds to times between 500 and 2000 ms respectively, or 1500 ms time frame. For the analysis, aimed at later use in real time biofeedback application, this is the most significant part of the swing. The signal recorded before the start of the movement is considered to be noise and signal after the impact is irrelevant for real time biofeedback as it occurs too late for action. Even more, for motor learning accelerating application the feedback should happen before the downswing phase, what is approximately 150 to 200 ms before impact.

Synchronized and segmented raw sensor signals are saved to the database. In addition to raw data, with each swing we also save a number of parameters defining the conditions and settings of the experiment at the time of the swing. Each swing is labelled with swing outcomes such as the distance, swing shape, player's evaluation, and many others. The application also allows reviewing and replaying the swing data from the database.

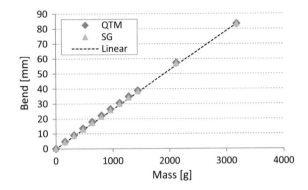

Fig. 7.12 Front strain gage sensor calibration and validation. The bend is measured directly by the QTM system and indirectly by the strain gage sensor. The horizontal axis shows the mass of the applied weight in [g], the vertical axis shows the bend of the club head in [mm] (Umek et al. 2017)

7.4.4 Application Testing

Application testing included laboratory and field tests. Static laboratory tests were primarily aimed at verification of the strain gage sensors in terms of their precision and for their calibration. Field tests were conducted to prove the correct operation of the application and to acquire a set of measurements for the smart golf club performance analysis and final sensor calibration.

Laboratory Test

Laboratory tests study the properties and the performance of strain gage sensors used for our smart golf club. Their final goal is the calibration of strain gage sensor readings and validation of their use for measuring the golf club bend.

A static test is performed with the club's grip firmly affixed to a massive table as shown in Fig. 7.8. The shaft bends for the deviation d under the known mass m that exerts the force $F = m \times g_0$ to the shaft just above the club's head. The bending deviation d is measured directly by the high precision reference QTM system.

The graph in Fig. 7.12 shows the bend of the golf club head in [mm] under applying the known mass exerting the known force $F = m \times g_0$. The bend is measured by the QTM system and is practically linear to the applied mass in the test range. The strain gage sensor readings are multiplied by the constant calibration factor to calculate the bend in [mm]. The constant calibration factor is determined from the slopes of the QTM and strain gage graphs. As seen from the graph in Fig. 7.12 the bend calculated from the strain gage readings is practically linear to the applied mass, thus we conclude that strain gage sensors are validated for measuring the bend of the golf club. For obvious reasons, IMU device was not used in static tests.

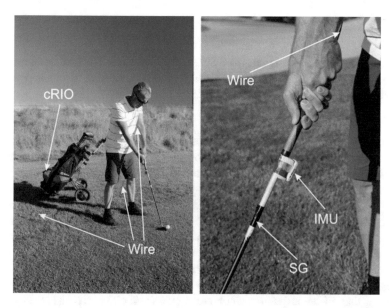

Fig. 7.13 Field test. The movement of the golf club during a typical golf swing is recorded by the SG sensors and the IMU device. SG sensors are connected by wire to the nearby cRIO device. The IMU device is connected wirelessly to the nearby laptop (not shown in pictures) (Umek et al. 2017)

Field Test

After the calibration and the validation of strain gage (SG) sensors in the static laboratory tests, we performed the pilot set of dynamic field tests to prove the correct operation of the experimental system elements.

Dynamic field tests include two SG sensors and IMU device, QTM system is not used. SG are connected by wire to the cRIO device, IMU is wirelessly connected to the laptop. cRIO and laptop are running the distributed LabVIEW application. As seen from the left hand side picture of Fig. 7.13, SG sensors are connected to cRIO by a thin wire. The IMU device is connected over a Bluetooth connection to the nearby laptop (not shown in Fig. 7.13). Laptop is used for communication with IMU for monitoring and controlling the measuring process. cRIO is running on batteries and is placed into the golf trolley. The exact placement of SG sensors and IMU device is seen in the right hand side picture of Fig. 7.13.

The dynamic test is performed by several experienced amateur golf players and two professional golf players. They consecutively perform a number of golf swings. In addition to that the professional golf players are asked to perform a number of swings with known and controlled swing execution errors.

7.4.5 *Results*

Field test results include signals from SG and IMU sensors. The first set of results is given to prove the measurement repeatability; the second set of results is dedicated to swing execution errors (technical errors).

Repeatability

The first set of swings was aimed to test the repeatability of sensor readings. For this purpose we have asked a professional golf player to carefully perform a series of consistent golf swings. We measured the response of two strain gage sensors, 3-axis accelerometer and 3-axis gyroscope. Sensor readings from the start of the swing to the impact of five swings are shown in Fig. 7.14. Sensor imprecision due to various factors (noise, bias drift, scaling factor drift, etc.) is negligibly small in comparison with effects of swing performance variations caused by the inconsistency of the player.

From the graphs in Fig. 7.14 can be concluded that all sensor signals during the series of consistent swings show high repeatability in both time and amplitude. It is expected that different features can be extracted from different sensors. While SG sensors measure the force exerted to the golf club and are closely correlated to the player's action, IMU sensors better reflect the golf club response.

Usability of Strain Gage Sensors

One of the aims of this paper is to examine the usability and suitability of strain gage sensors for the golf swing analysis. Figure 7.15 shows the readings of the front and side strain gage sensor of five swings of four different golf players.

From the graphs in Fig. 7.15 we can conclude, that considering the findings about the repeatability, all players performed consistent swings. The swing consistency is also evident from the experiment records meta data where all the players subjectively labelled the swings as correctly executed.

Apart from the repeatability and consistency, the graphs in Fig. 7.15 show that all tested players have their characteristic and distinctive signal shapes – player's signature. For example, the player in the graph in Fig. 7.15b starts the swing at approximately 600 ms, while other players start the swing at approximately 200 and 400 ms. Similarly we can observe that players in graphs Fig. 7.15a, c have a bounce in the red signal just before the impact, while the other two do not. Other similar examples can be derived from Fig. 7.15.

Even the top professional golf players have different golf swings. Their golf swings depend on their body constitution, their physical abilities, and their personal style. The objective of golfers in learning or training is not to repeat the swing of another player, but to learn to repeat their best swing; for example, the swing that results in the intended ball flight. Each golfer should strive to be able to repeat the good swing over and over again.

A more intuitive representation of the difference in players' signatures is given in Fig. 7.16. The 2D strain graph is created by plotting the readings of the side SG

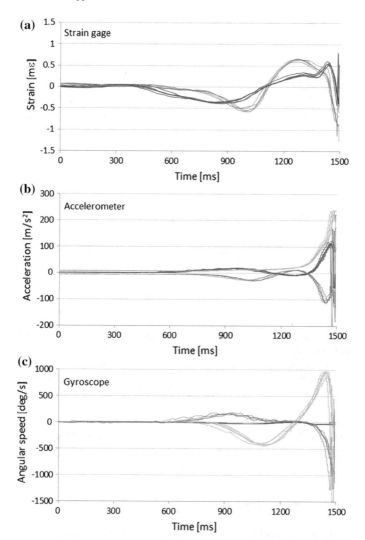

Fig. 7.14 Measurement repeatability test. Sensor signals for the series of five swings performed by a professional golf player: **a** two orthogonal SG sensors; **b** 3-axis accelerometer; **c** 3-axis gyroscope. In graph **a** blue plots represent the front sensor and red plots represent the side sensor, in graphs **b** and **c** red, green, and blue lines represent X, Y, and Z axis respectively (Umek et al. 2017)

sensor to the horizontal axis and the readings of the front SG sensor to the vertical axis of the graphs. It is obvious that the shape of the plots does not depend on the quality of the player. Based on the observed plots, we expect that the identification of a player is possible with a high probability. The player identification is not one of the main functionalities of this application; however the results in Figs. 7.15 and 7.16 can give us important directions for the application design. It is important to

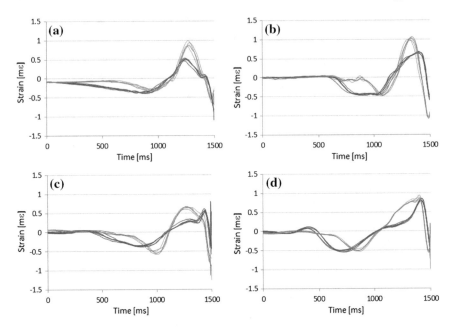

Fig. 7.15 Player's signatures of five successful consecutive swings of four different golf players: graphs **a** and **b** show swings of two experienced amateur players; graphs **c** and **d** show swings of two professional golf players. All graphs show the response of two orthogonally placed SG sensors, where blue plots represent the front sensor and red plots represent the side sensor. Signals show that each player has a distinctive sensor response (signature) (Umek et al. 2017)

consider the fact that correctly performed swings of different users are different. The consequence of this observation is the need for personalization of the swing error detection application; the application learning phase will have to be performed separately for each player.

Detection of Golf Swing Technical Errors

One of the primary goals of every golf swing analysis is the detection of technical errors made during the swing execution. Through the identification and elimination of error causes players can improve their swings and consequently advance in mastering golf.

Traditionally, error analysis is done after the golf swing, in post processing, giving players a terminal feedback. The goal of this application is to gives players concurrent biofeedback. That means that a player is notified about possible technical error in swing execution as soon as the error can be detected; if possible, before the downswing phase.

All graphs in Fig. 7.17 show that the swing with slice technical error can be distinguished from the correct swing. With accelerometer we observe slight deviation of the timing in the X and Y axis around 1000 ms mark, and the amplitude in the X axis just before the impact. Similarly, gyroscope shows timing deviations in Y

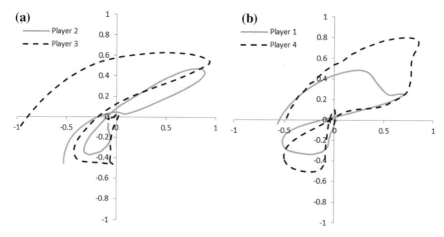

Fig. 7.16 Player's signatures shown in 2D SG plot indicate that each player has a very distinctive sensor response. Graph **a** shows signatures of two experienced amateur players; graph **b** shows signatures of two professional golf players. The strain in both axes is given in [me]. The horizontal axis shows the strain of the front SG sensor and the vertical axis shows the strain of the side SG sensor (Umek et al. 2017)

and Z axis, while amplitudes are practically the same for both series of swings. The greatest difference in signal traces is observed with strain gage sensors. Especially important is the amplitude difference at approximately 1250 ms mark that represents the top of the backswing. That points out, that from all of the sensors used in our experiment; strain gages are probably the most suitable for the detection of (slice) technical errors.

Another observation is that when comparing all the graphs in Fig. 7.17, the strain gage sensors are the first that show any response (signals different from zero value) to the performed movement. The reason for this is that they represent the force applied to the golf club, while accelerometers and gyroscopes represent the response (the movement of the golf club). This observation supports the claim that strain gage sensors, being closer to the cause of the error, can be more suitable for technical error detection. Another advantage, especially for the concurrent biofeedback, is that strain gage sensors show the error earlier in time and therefore allow enough time for possible swing interruption (Umek et al. 2015).

To explore further, we plot three different swing shapes: straight, slice, and draw. 2D strain graphs for two professional golf players are given in Fig. 7.18. As expected, shapes of the same swing type differ greatly between the players. This is consistent with the observations from Fig. 7.16, where each player has very distinctive swing signature. From both graphs in Fig. 7.18 it can be observed that trajectories depend on swing shape, but the difference between them is not as great as the difference between players' signatures in Fig. 7.16. Based on the plots in Fig. 7.18, we expect that the identification of different technical errors, within the swings of the same player, is possible.

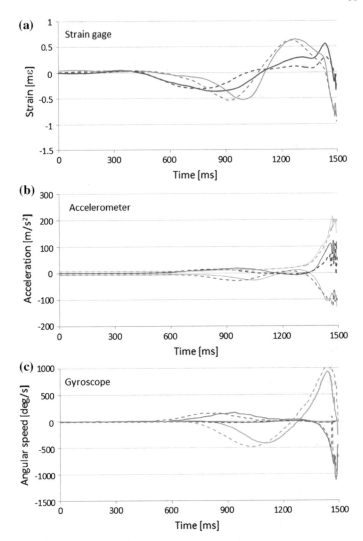

Fig. 7.17 Technical error detection. Averages of five consecutive swings: **a** two orthogonal strain gage sensors; **b** 3-axis accelerometer; **c** 3-axis gyroscope. Solid lines represent averages of successful swings; dashed lines represent averages of swings with "slice" technical error. Traces of the same colour are distinctively different from each other. We expect that technical error detection is possible. In graph **a** blue plots represent the front sensor and red plots represent the side sensor, while graphs (**b**) and **c** red, green, and blue lines represent X, Y, and Z axis respectively (Umek et al. 2017)

Technical Error Detection Results

Detection of swing type and swing technical errors in this section is based on signal correlation coefficient comparison. Swing data from K swings is grouped by player and swing type $G(Player, SwingType)$. Metadata $Player$ and $SwingType$ is a priori

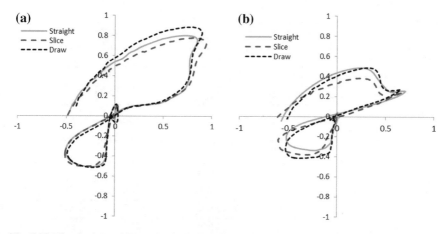

Fig. 7.18 Comparison of 2D strain plot for regular swings and swings with technical errors of two professional golf players: **a** Player 1; **b** Player 4. Regular swings (straight) differ from swings with technical errors (slice, draw) in one or more swing phases. We expect the identification of technical errors is possible. The strain in both axes is given in [me]. The horizontal axis shows the strain of the front sensor and the vertical axis shows the strain of the side sensor (Umek et al. 2017)

Table 7.2 Probability of swing error detection based on signals of different sensors (Umek et al. 2017)

Swing type	Detection accuracy		
	Strain gage	Accelerometer	Gyroscope
Straight	0.93	0.69	0.72
Slice	1.00	1.00	1.00
Draw	0.83	0.67	0.83

knowledge, available for each swing. Each group of swings is characterized by a normalized average sensor signals, which serve as a reference for comparison with all analysed swings. Each signal is compared to all group reference signals by calculating the Pearson correlation coefficients $C(k,G)$. Detection of the most probable swing type is made according to the max-correlation coefficient criteria. The same procedure is used for all sensor signals.

The results of the technical error detection of the first field tests with a professional golf player are listed in Table 7.2. Probability of a successful golf swing technical error detection based on signals of different sensors is approximated with a relative frequency of an event. The total number of swings performed by Player 1 is $K=67$. The detection accuracy achieved by strain gage sensors is at least as good as the accuracy achieved by the accelerometer or the gyroscope.

Swing error detection accuracy can vary by players. The test results for a mixed group tests with two players and three golf swing errors are presented in Table 7.4, while experimental settings parameters are listed in Table 7.3. The number of analysed swings is $K=84$ that include three of the nine swing types: straight, slice, and

Table 7.3 Experimental setting parameters (Umek et al. 2017)

Parameter	Notation	Value
Analysed swings	K	84
Signal samples for error detection	N_{imp}	750
Signal samples for concurrent feedback	N_{FB}	625
Players		2
Different swing types		3

Table 7.4 Probability of swing error detection from strain gage signals (Umek et al. 2017)

		Detection accuracy	
Player ID	Swing type	750 samples	625 samples
1	Straight	0.93	0.88
1	Slice	1.00	1.00
1	Draw	0.83	0.83
2	Straight	0.91	0.73
2	Slice	1.00	0.50
2	Draw	0.75	0.75

draw. The detection accuracy is calculated by taking into account all signal samples from the takeaway (golf swing start) to the impact, when the club hits the ball (N_{imp}=750). It is known (Umek et al. 2015) that the concurrent feedback in golf swing training should be offered before the downswing phase begins. That is approximately at the 625th sample (N_{FB}). For this reason it would be very useful to detect a golf swing technical error during the backswing, that is, in the first 625 samples. Results in the last column of Table 7.4 show that early detection accuracy is lower than the detection accuracy with signal samples up to the impact. The ranking of the results in Table 7.4 can be compared by grading the 2D SG plots similarity in Fig. 7.18. For both players it is possible to claim that the slice swing error in Fig. 7.18 differs the most and that the measured detection accuracy for a slice swing error is expected to be higher than for other swing errors.

7.4.6 Future Development

This application proves the usability of strain gage sensors for golf swing analysis and compare them to more widely used inertial sensors; for the analysis of the golf swing they complement each other. Strain gage sensors are especially suitable because bending of the club shaft reveals some information of the swing performance which could not be obtained from inertial sensors, which, due to range limitations, cannot be placed on club head.

For the concurrent biomechanical feedback application, the exact orientation, velocity and position of the club head are not of the major importance. The most important is the "signature" of the proper swing of each individual player, i.e., the swing that the golfer would like to repeat again and again.

The idea is to detect any deviation from this "signature" and "warn" the golfer to suspend the swing in its early phase, as soon as the error is detected. Ideally already at the beginning of the backswing, so that the player does not "learn" improper motion. Experiments showed that improper motion or even improper setup could be easily detected from the bending of the club shaft, easier than from data from inertial sensors placed on the hand of the golfer or at the upper part of the shaft. The results gained by a basic cross-correlation signal analysis, and presented in Tables 7.2 and 7.4, confirm our presumption and show that the detection accuracy achieved by strain gage sensors is at least as good as the accuracy achieved by the accelerometer or the gyroscope. We expect to get more reliable and more detailed results by using more sophisticated data analysis methods on a larger data set, such as presented in Guo et al. (2016), Jiao et al. (2018), Li et al. (2017), Sakurai et al. (2016), Wei et al. (2016), and Zhang et al. (2017).

Our idea is to develop a much more demanding application that would be capable of immediate detection of errors and giving appropriate concurrent biofeedback to the player. This includes the development of the system for real time swing analysis and the design of mobile and cloud applications with terminal or concurrent feedback (Sun et al. 2016). Such applications will enable golf players to speed-up motor learning in golf.

7.5 Smart Ski Application

Similarly to the *Smart golf club* application presented in Sect. 7.4, *Smart ski* is an application based on smart sport equipment. Smart ski is an example of a biofeedback system with concurrent, cyclic or terminal biofeedback based on smart sport equipment. The application is aimed at helping skiers to learn or improve their skiing technique.

7.5.1 Objectives and Functionality

The smart ski application is designed for user and instructor functionality. It has a compact structure and can be used in personal or open space. The smart ski application is intended for all levels of skiers: beginners, intermediates, experts, and professionals.

- Beginners may use the application to learn skiing in the concurrent or cyclic biofeedback mode through a set of predefined lessons and courses supported by the Smart ski application.
- Intermediate and expert skiers may use the application for monitoring their skiing technique during the skiing action in the cyclic or terminal biofeedback mode. For example, skiers can follow their carving turn correctness through parameters given by the application right after each turn completion (cyclic feedback). All the other skiing parameters can be reviewed after the skiing action; immediately at the bottom of the slope or at any time later (terminal feedback).
- Professionals may use the application for monitoring the imperceptible changes in their skiing technique. For example, when the ski slope allows, downhill competitors strive to drive skis as flat as possible to the surface; smart ski application can provide information about the contact of the ski with the surface: on the left edge, flat, and on the right edge.

The smart ski application is aimed primarily to help skiers and trainers to better understand the reaction of the skis to the applied skier's action and consequently help skiers to accelerate motor learning or improve their skiing technique.

The application measures and calculates the force applied to the skis, the resulting bend of the skis, and the motion parameters of the skier (acceleration and angular speed).

7.5.2 Background

Wearable devices and smart sport equipment are slowly but surely bringing the technology into sport, recreation, and our daily life (Ebling 2016). In general, science and technology are increasingly used to augment sport training and exercise (Kunze et al. 2017). Smart sport equipment is a sort of a smart system. To achieve a goal, smart systems employ sensing, processing, and actuating devices for smart actions that are adaptive and predictive.

Skiing is a very complex motion that depends on many parameters and conditions; from the snow and weather conditions, to the type and model of skis, from the skiers' skills to their physical state. Many physical quantities must be monitored in order to accurately capture and analyse skiing: speed, skier's position, terrain inclination, acceleration, ski bend, applied force, and others. A number of different sensors are needed for their acquisition. The smart ski application employs several different sensors. A detailed analysis of skiing motion requires sensor fusion that combines data from different sensors in order to lower the uncertainty of information from separate sensors. For example, the detection of a ski carving motion is more reliable by adding an accelerometer to edge load sensors.

Skiing has not been studied as extensively as some other sports. One of the reasons is probably the high complexity of sensing, monitoring and analysis of skiing motion. Some works that shed light on the subject are studying various parts of the problem

(Adelsberger et al. 2014; Falda-Buscaiot et al. 2016; Kirby 2009; Michahelles and Schiele 2005; Nakazato et al. 2011; Nemec et al. 2014; Umek and Kos 2018a,b; Yu et al. 2016). In the most works post processing is used and different combinations of sensors and sensing systems are employed. Adelsberger et al. (2014) concentrate on the bending characteristics of the skies during motion, reaction forces of the skier are the core of the work in Nakazato et al. (2011), in Yu et al. (2016) the stress is in studying the potential of IMU (Inertial Motion Unit), Nemec et al. (2014) use machine learning to get the approximation of the exact motion of a skier, and a real-time biofeedback system is presented in Kirby (2009).

7.5.3 System Architecture and Setup

The system consists of sensors integrated into the skis and attached to the skier; it consists of measuring devices for sensor signal acquisition, processing devices for sensor data analysis, and feedback devices for communicating the results back to the skier. Laboratory tests for equipment operation testing, calibration and validation were followed by several snow tests in different weather and snow conditions and performed with different expert skiers.

Terminology

For the understanding of the system architecture, results, and explanations given later in this section, some basic denotations and skiing related terms are given first. Figure 7.19a shows a pair of left and right ski with their designated inner and outer edges. Figure 7.19b presents the situation in the left and right turn during skiing. For example, as shown in Fig. 7.19b, in the left turn the left ski is the inner ski of the turn; the outer edge of the inner ski and the inner edge of the outer ski are applied to perform the turn.

One of the important skiing parameters is the correlation between the force applied to the skis and the resulting ski bend that is graphically shown in Fig. 7.20. The geometry of the ski is slightly lifted in the ski centre; when the ski is unloaded and lying flat on the supporting points, its centre is slightly higher than the supporting level. We denote the level of the centre of the unloaded ski as *zero level* and the level of the supporting points as *ground level*, see Fig. 7.20. By applying force F to the centre of the ski, it bends and its centre is displaced for the Dh to the *flexing level*. The ski bend is an important parameter because it defines the radius of the carving turn.

SmartSki Prototype

The SmartSki prototype is designed to measure the force applied to the skis by the skier and the corresponding bend of the skis as presented in Fig. 7.20. For that purpose we have installed several force sensors and bend sensors to different locations on each ski, as shown in Fig. 7.21.

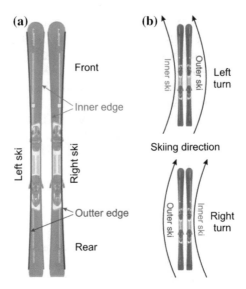

Fig. 7.19 Denotation of **a** skis, ski sections, ski elements, ski components and **b** ski positions during skiing regarding the direction of skiing (Kos and Umek 2018)

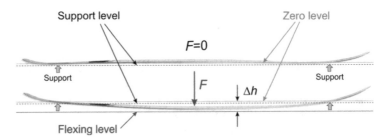

Fig. 7.20 Ski bend is measured by applying the force F to the point equidistant from both support points. The middle of the unloaded ski at $F=0$ is denoted by *zero level*, which is normally above the *ground level* defined by the supporting points. After applying the force F the ski bends for Dh (Kos and Umek 2018)

During skiing, it is very important to know the distribution of the force applied from the skier to each ski. The SmartSki measures the applied force in four points under each ski boot, two at the toes and two at the hill of the ski boot; in both locations sensors are placed at the left and the right edge under the ski boot. Therefore it is possible to determine the general distribution of force. For example, it is possible to simultaneously determine the distribution of force in the front-rear direction and in the left-right direction for each ski separately.

Bend sensors are positioned at the front and the rear of each ski as shown in Fig. 7.21. The SmartSki can measure the bend separately for the front and for the rear section of each ski. It is possible to get the correlation between the applied force and the bend of each section. For example, when the skier's force is applied primarily

Electronics   Force sensors

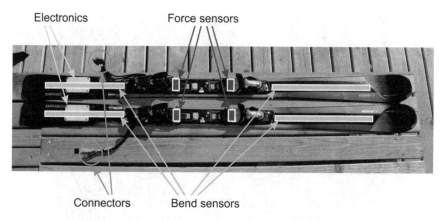

Connectors   Bend sensors

Fig. 7.21 SmartSki prototype with marked sensors for measuring the bend of each ski, sensors for measuring forces applied to each ski, wires with connectors, and mounted box with electronics and sensor interconnections. Ski bend is measured separately at the front and rear section of each ski, force is measured in four points under each ski boot (Kos and Umek 2018)

to the front (the toes of the skiing boot) the front section of ski bends stronger than the rear section of the ski.

All sensors are connected to the measuring equipment by wires. The electronics box mounted at the rear of the skis serves as a central interconnection for the sensors and as the place for electronics needed to drive the sensors. Each ski connects to the measuring equipment by three 8-wire cables with connectors, as shown in Fig. 7.21.

Sensors and Processing Devices

Sport equipment integrated sensors should not interfere with the sport activity itself; therefore they must be small-size and lightweight. They should not change the performance of the equipment and they should not physically obstruct the activity. While the SmartSki is not comfortable (wires, backpack), it does not physically obstruct the skiing actions of experts on the performed tests.

The system includes bend sensors for measuring the bend of the ski in several sections, force sensors for measuring the force that skier is applying to the ski, 3-axis accelerometer, and 3-axis gyroscope. Bend and force sensors are integrated into the ski, accelerometer and gyroscope are attached to the skier's torso. Sensor signals are collected, synchronized and processed by the LabVIEW™ application running on cRIO platform from National Instruments. The sampling frequency of the system is 100 Hz.

Smart ski application consists of professional measurement equipment stored in the skier's backpack, the SmartSki, an IMU device mounted at the skier's torso, a high definition GoPro™ Hero camera, and a laptop (see Fig. 7.22). The heart of the system is the professional measurement equipment cRIO from National Instruments. It is connected to the SmartSki prototype by wires and wirelessly over the WiFi connection to the IMU device and to the laptop. cRIO is a standalone system that can

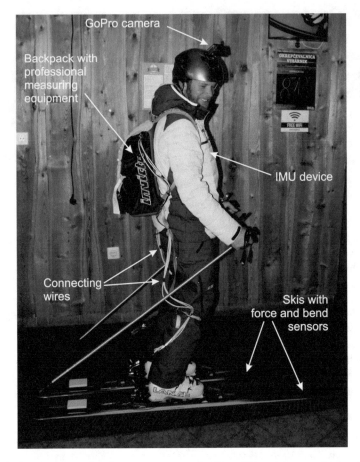

Fig. 7.22 Development and testing system composed of SmartSki prototype, professional measuring equipment, and IMU device, GoPro™ Hero camera (Kos and Umek 2018)

work in a real-time mode without connection to the laptop, which is used primarily for terminal feedback, monitoring the results, and controlling the system's settings. cRIO runs the LabVIEW™ application that gives the skier concurrent or cyclic feedback.

High definition GoPro™ Hero camera is not directly connected to the cRIO or laptop, its video is synchronized with other signals in post processing if needed. The dynamic range of the accelerometer is $\pm 2g_0$; the dynamic range of the gyroscope is ± 2000 deg/s.

Fig. 7.23 System calibration: **a** full calibration of SmartSki and skier parameter setting just before snow test; **b** quick calibration of SmartSki during the snow test, just before each individual test run

7.5.4 Application Tests

A series of snow tests were performed, which included several different expert skiers; former professional *FIS race* skiers and members of the *Slovenian Alpine Demo Team*. Snow tests were performed on different slopes in different skiing resorts over a period of one year. The system presented in Sect. 7.5.3 is aimed to precise measurements; therefore a lot of careful preparations and a number of tasks have to be performed before and during each snow test.

Calibration

Calibration of sensors is of high importance for quality measurements. Before each snow test a series of tasks is performed inside the building on a flat surface. All sensors are thoroughly tested and calibrated, skier's parameters, such as weight, are entered into the system, etc. The initial calibration procedure is shown in Fig. 7.23a. During the snow test and before each individual test run, a quick calibration is performed to compensate for possible environmental effects, such as temperature changes, on the sensors and consequently measuring results, as shown in Fig. 7.23b.

Test Runs

All test runs were performed on a slope with medium inclination that allows skiing in wide or narrow corridor and gives the expert skier a perfect control over skis and a wide range of possibilities in performing various skiing techniques. The majority of performed tests were free runs where the test skier had the freedom to choose the skiing corridor according to the predefined task, as seen in Fig. 7.24.

In test runs expert skiers performed various skiing tasks and techniques according to the predefined schedule. For example, one of the tasks was to perform the carving turns by equally loading both skis, another task was to perform perfect carving turns, yet another task was to ski as dynamically as possible, etc.

Fig. 7.24 Free run test performed on a slope with medium inclination. The test skier is performing the predefined task – balanced carving technique

Fig. 7.25 Test run example: right carving turn graphically augmented with its radiuses and inclinations of the skis to the terrain (Kos and Umek 2018)

A regular right carving turn, graphically augmented with some of its features, is shown in Fig. 7.25. For example, the radius of the carving turn (without sliding) can be directly measured by the application. The bend of the ski is one of the quantities measured and it is inversely proportional to the carving turn radius. Other quantities, such as the distribution of force applied to the skis, skier's torso acceleration and angular speed, are also measured and can be combined to derive other important skiing features.

Our goal is to use the measured and derived quantities of the system in applications that help skiers to learn skiing faster and more properly, trainers to identify minor errors in skiing technique, ski teachers to give learners the predefined tasks to follow, ski manufacturers to monitor ski performance, and ski rentals to help advise the customers of the most appropriate equipment.

7.5.5 User Interfaces

Development and Instructor Interface

The development and instructor interface is primarily aimed at skiing analysis. It combines the sensor data with the synchronized video acquired by the GoPro™ Hero camera mounted onto the skier's helmet. One of the application's analysis windows is shown in Fig. 7.26. Such analysis helped us define certain parameters and check some others. For example, by inspecting the video and bar indicators for ski bending level we can check the correlation between the observed bending and bending measured by integrated sensors.

The development and instructor interface gives very detailed feedback about performed skiing actions. Its main purpose is to monitor skiers' actions for the Smart-Ski application development (developers) and for detailed insight about the skiing technique (instructors, trainers). If the information about the skiing performance is forwarded back to the skier, then the application can be classified as a biofeedback application with terminal feedback. This user interface is available only in post processing

User Interface with Concurrent and Cyclic Biofeedback

The user interface provides the skier with concurrent or cyclic feedback. For example, the skier is receiving the feedback information of the skiing performance during the turn or after each turn is completed.

The feedback can be given through different modalities. The most common are visual and auditory feedback. In our application the visual feedback uses graphical representation of skiing performance that should be simple and easy to understand. It is also important that it is non-distracting to the user and that its cognitive load is relatively low. An example of a concurrent visual feedback, projected onto the skier's goggles is shown in Fig. 7.27. An exemplary display including binary state indicators (carving, crossing, on the edge) and sliders showing skiers current performance (outer/inner, front/rear) is shown in Fig. 7.27. Auditory feedback uses different sound techniques for presenting feedback information. For example, the information about the inner/outer balance performance could be given as a stereo sound that is perceived by the user as moving to the inner or outer ski direction, according to the slider movement in Fig. 7.27.

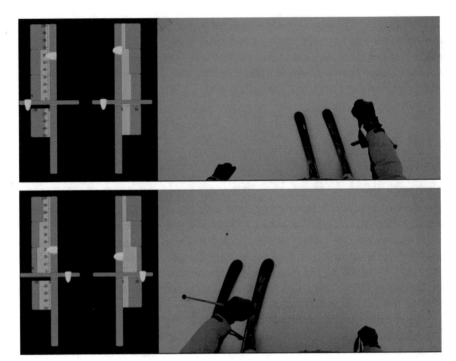

Fig. 7.26 Sensors signals follow the skier's action: vertical green bar sliders represent the weight distribution in the front-rear direction, horizontal green bar sliders represent the weight distribution in the left-right direction, and bar indicators represent the bending level in four positions along each ski (Kos and Umek 2018)

7.5.6 Results

Smart ski application produces large amounts of data. Only a few examples of raw sensor signals and a few examples of processed and derived signals are presented here. Figure 7.28 shows a selected group of raw signals from different sensors. The ski run presented in Figs. 7.28 and 7.29 consist of seven high dynamic double turns between $t = 51$ s and $t = 73$ s.

Raw sensor signals are noisy due to the ground reaction force vibrations generated by various type of snow terrain roughness. Terrain vibration noise is detected by the highest amplitude on bend sensors that are installed closer to the noise source. Figure 7.28a shows signals from bend sensors positioned behind the rear bindings on both skis. Noise is also visible on force sensor signals in Fig. 7.28b and even on the accelerometer positioned on the skier's body in Fig. 7.28d. The high frequency noise with high amplitude is usual on a hard frozen snow and can be reduced by a low-pass filter.

Figure 7.29 shows processed sensor signals from the same ski run. It is magnified in a narrower time interval. A series of four high dynamic turns is shown, starting

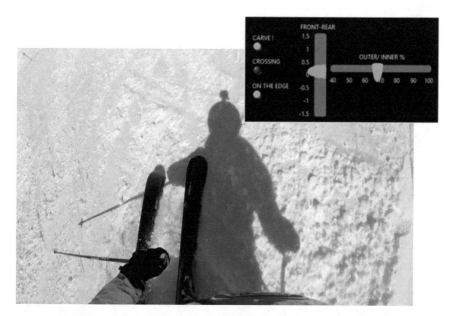

Fig. 7.27 Concurrent feedback can be presented to user visually by a snow googles with a display. Concurrent data presentation of the visual feedback information, shown on the head-up display of the googles, depends on a planned skiing lesson and the skiers' performance (Kos and Umek 2018)

with a right turn at $t = 62.5$ s. Bend signals in Fig. 7.29a confirm that the outer ski experience higher load and consequently its bend is larger comparing to the inner ski. The total load shown in Fig. 7.29b is calculated from filtered force sensor signals. The ski edge changing phase is most clearly visible from the edge load distribution graphs in Fig. 7.29c, where the edge signal represents the sum of all force sensor signal applied to the same edge for both skis. The steering phase occurs during the high load, where the ski is on the edge and the skier is turning in the direction of the high loaded edge. The turn direction can be observed from the edge balance signal in Fig. 7.29d, which is derived from signals in Fig. 7.29c, b as a normalized edge load difference. The skier started each turn with the anterior movement of the centre of gravity, followed by the posterior movement. The latter is observed from the derived front/rear force distribution signal in Fig. 7.29e.

The meaning of the measured sensor signals can be confirmed by comparing them with the synchronized video records in post processing. Processed sensors' signals can be presented to sport experts in more conceivable way as shown in Fig. 7.26. The pair of crossed slider indicators presents the load force distribution of both skis: green pointer on horizontal slider measures the ski load distribution between both edges. Similarly, the green pointer at vertical slider measures the load distribution along the ski. Ski bending, measured by bend sensors, can be observed from bar indicators positioned at the outer side of vertical load indicators.

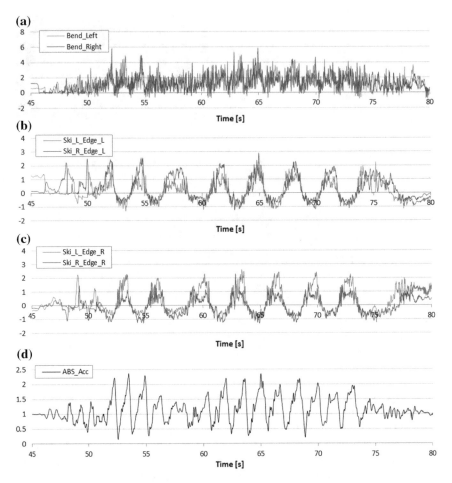

Fig. 7.28 Raw signals from different type of sensors: **a** signals from both ski equipositioned bend sensors, **b** combined force sensor signals on left edges, **c** combined force sensor signals on right edges, and **d** acceleration vector magnitude (Kos and Umek 2018)

The skier spends less energy and less time to control the edges simply by moving the sole of the foot than to move the centre of gravity from backward to forward position or vice versa. As a consequence the edge distribution can change much faster than the rear/front distribution, as can be observed from Fig. 7.29d, e.

An example of concurrent feedback signal presentation is illustrated in Fig. 7.27b. In this case the skier is training the lesson that requires the focus on driving the perfect carving turns. The feedback information includes the inner/outer ski load balance, forward/backward balance and indicators with automatic detection of the performed ski movement (carving, crossing, on the edge).

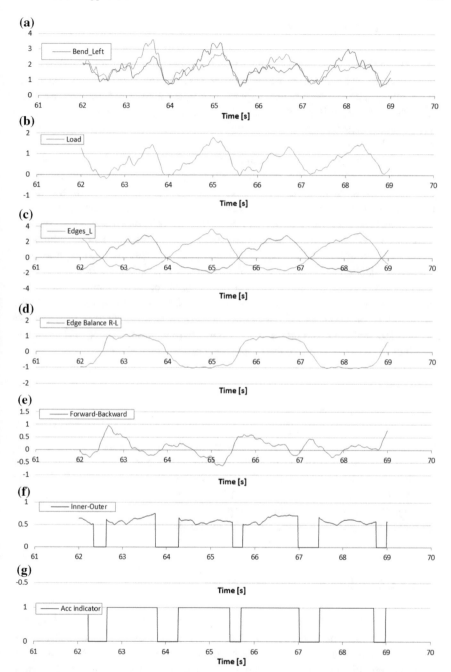

Fig. 7.29 Processed sensor signals: **a** filtered ski bend signals, **b** filtered sum of force sensor signals represents a total dynamic load, **c** filtered sum of force sensors on left and right ski edges, **d** edge balance = normalised difference between right edge and left edge signals, **e** tilt balance = normalised difference between front and rear force signals, **f** relative load of outer ski, and **g** indicator of turn loading phase based on acceleration magnitude (Kos and Umek 2018)

7.5.7   Future Development

Smart ski application can be used for various purposes; from accelerated motor learning and skiing technique improvement of the skier, to measuring skis performance for the manufacturers and other ski experts. Our vision is to truly combine the Smart-Ski capabilities and skiing experts knowledge to devise a set of exercises tailored to specific skiing parameters that would be performed through a concurrent or cyclic biofeedback application. For example, many beginner and intermediate skiers have considerate problems with the tilt balance; see Fig. 7.29e. Smart ski application with concurrent biofeedback would monitor the parameter in question and give visual or audio feedback to users about their performance or when predefined tilt thresholds are exceeded, an alarm would go off. Many similar simple exercises based on one or more skiing parameters and backed by concurrent or cyclic biofeedback are possible.

Our vision is also to make SmartSki a part of the Internet of Things, which can offer almost endless possibilities. For example, skiers could use cloud applications to monitor their progress made by following the exercise course developed for a mobile biofeedback application. Manufacturers could get the information about skis' performance and stress in real environment. Professionals could measure parameters that cannot be observed by the eye of the trainer or by traditional video analysis. We are confident that SmartSki offers many benefits to recreational skiers, ski equipment manufacturers, ski schools, coaches, and even professional skiers.

7.6   Water Sports

Biofeedback applications in water sports are specific because the activity is being carried out in the water or in the water environment. Consequently all the devices used in biofeedback systems and applications for water sports must be water resistant or waterproof. Three sports are included in this section: swimming, kayaking, and canoeing. The presented biofeedback applications use the combination of body and equipment mounted sensors and they give cyclic and terminal feedback.

7.6.1   Objectives and Functionality

Water sport biofeedback applications are designed primarily for professional athletes and their coaches. The water environment is limiting the normal use of many of the devices used in other sport related biofeedback applications. For that reason the presented applications focus primarily on the instructor functionality of the system and less on the user functionality, as defined in Sect. 4.4. Applications are used in open space and have a distributed structure.

The three applications for swimming, kayaking, and canoeing have some elements and functionalities that are common to all of them and some that are specific to each.

- *Common elements and functionalities* are similar cyclic movements called strokes. In swimming strokes are made by arms (and leg kicks), in kayaking with a double bladed paddle, and in canoeing with single bladed paddle.
- In *swimming* sensors are attached to the body of the user at the centre of mass at the lower back. Sensors must be waterproof.
- In *kayaking* sensors are attached to the kayak and to the paddle. Attachment to the user is not necessary as the body does not move significantly. Sensors must be water resistant.
- In *canoeing* sensors are attached to the body of the user at the centre of mass at the lower back, to the canoe, and to the paddle. Sensors must be water resistant.

Water sport applications are intended primarily for coaches who give the athletes the terminal feedback. Cyclic feedback on carefully selected parameters can also be given directly to the athlete. For example, stroke symmetry or stroke timing variations.

7.6.2 Background

Use of various sensors, tools, and technology in swimming has received a lot of research attention. Tools like hand paddles (Barbosa et al. 2013) and mechanical measuring systems (Dopsaj et al. 2003) have been extensively used. Recently the most popular is the use of inertial sensors attached to the swimmer's body. The most comprehensive review can be found in Mooney et al. (2015), where 87 articles are systematically analysed and classified by stroke style, sensor type and technical parameters, data storage and communication, swimming output variables and validation methods. The most popular sensor body attachment points are the lower back, wrist and lower arm and the most prevalent sensors are accelerometers and gyroscopes. A shorter review in Magalhaes et al. (2015) offers a view on the same topic. A selected group of 27 articles was analysed and classified by sensor type and specification, body area, stroke style, swimming variables, validation method and data analysis method. The most popular sensor placement for motion monitoring in swimming are found to be the lower back, the head, wrist, and ankle. The majority of earlier research work uses only accelerometers. One of the main reasons is that the MEMS gyroscopes become available on the market a few years later than accelerometers. Most modern IMU devices have integrated both sensors inside the same microchip.

Swimming is characterized by a sequence of coordinated actions of the trunk and limbs, in a repeated, synchronous pattern Mooney et al. (2015). We can expect that at ideally performed free swimming, with perfect synchronous and periodic motion, all body attached sensor signals should be periodic. The period is related to stroke time. In practice this signals are cyclic with small differences in period length and period signal shape. The variation of stroke period can help coaches to grade the swimming action consistency. Similar specifics in sensor signals can be found in somehow familiar water sports as rowing (Llosa et al. 2009; Tessendorf et al. 2011), kayaking (Sturm et al. 2010), and canoeing (Wang et al. 2016).

Stroke periods and sport specific signal peak values are listed as the key performance indicators in many water sports. The most important parameters and tasks for free swimming analysis are found to be (Mooney et al. 2015): stroke phase analysis, stroke type identification, lap time, swim distance, stroke count and rate period, swimming velocity, kick count and kick rate. Various other parameters are important in analysis of starts and turns. Our research work focuses on the acquisition of as many key performance indicators as possible by using a simple and robust measurement system that can be adopted by coaches without any need for specialized technical support. Most coaches are still using only the direct visual information and stopwatch. We are deeply convinced that a relatively simple and inexpensive technical support system can give them very valuable supplemental information.

The swimming application uses a single 6DoF IMU device. Similar sensor type and setup can be found in Stamm et al. (2013). We selected the lower back as the most appropriate placement for the sensor unit. Application test results show that a single 6DoFIMU device is sufficient for acquiring a number of important swimming parameters, such as accurate stroke counting and stroke rate measurements. Other researchers used accelerometer signals with algorithms, which involve the detection and summation of acceleration peaks (Ganzevles et al. 2017). The application uses rotation angle signals around the principal axis, which is the most relevant for specific swimming discipline.

Similarly, the application uses only 6DoF IMU devices in kayaking and canoeing, but the number of sensor nodes was increased due to additional sport requisites (kayak/canoe, paddles). The number of sensor nodes in kayaking and canoeing is not as disturbing factor as in swimming and can be further increased with various sensor types. The application uses similar method for stroke counting and stroke rate measurement as in swimming; the only difference is in the cycling signal, which provides enough information for stroke analysis. In kayaking and canoeing we found the boat moving direction acceleration signal as the best choice.

7.6.3 System Architecture and Setup

The application uses several IMU devices with 6DoF MEMS sensors. Each device includes 3D accelerometer and 3D gyroscope. Two IMU devices are fully waterproof; the other two are water resistant. The application uses different number of selected IMU devices in each sport, according to the needs of each.

Swimming

In swimming the application uses only one waterproof IMU device. Figure 7.30a shows the position of motion sensors attachment on the swimmer. We have used a

(a) **(b)**

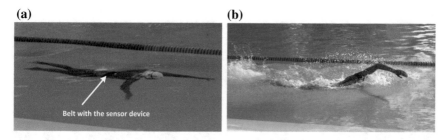

Fig. 7.30 Swimming: the movement of the body is detected by a sensor attached to the belt, which is strapped to the lower back of the swimmer. The belt is visible in the underwater gliding phase photo (**a**). The belt and the sensor do not interfere with the athlete during the swimming action (**b**)

(a) **(b)**

Fig. 7.31 Kayaking: one waterproof sensors is attached to the top of the kayak (**a**), two water resistant sensors are attached to the paddle (**b**)

custom designed belt that held the IMU device firmly in place of attachment even in power swimming, as seen in Fig. 7.30b.

Kayaking

The kayaking application uses three IMU devices. The waterproof IMU device attached to the kayak and two water resistant IMU devices attached to the paddle, one at each arm, as seen in Fig. 7.31a, b.

Canoeing

The canoeing application uses three IMU devices. The waterproof IMU device attached to the canoe, one water resistant IMU device attached to the paddle, and the other water resistant IMU device strapped to the athlete inside the custom designed belt, as seen in Fig. 7.32. Kayaking and canoeing actions were monitored by a water-proof GoPro camera, as seen in Fig. 7.33.

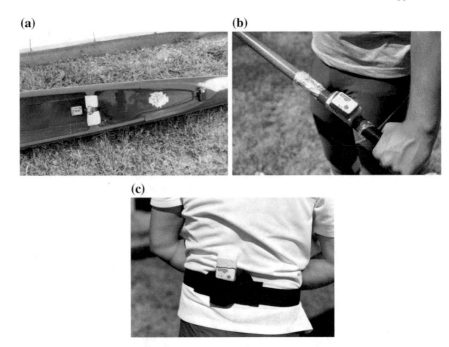

Fig. 7.32 Canoeing: waterproof sensor devices is attached to the bottom of the canoe (**a**), water existent devices are attached to the paddle (**b**), and inserted into the belt (**c**)

Fig. 7.33 Athletes in kayaking and canoeing are monitored by a GoPro camera from different view angles: **a** kayaking is monitored from the rear; **b** canoeing is monitored from the front

7.6.4 Results

The water sports application is in the final development phase. The field test results presented in this section give valuable information to the developers and coaches about the parameters that can be measured and calculated for each sport. Based on this information and the feedback needs in each of the sports, the final application version will be prepared.

Swimming

Swimming test consist of a single continuous record of 12 laps along a 50 m swimming pool with intermediate interruptions for resting. The complete record of accelerometer and gyroscope signals from the IMU device is shown in Fig. 7.34.

A sequence of twelve single lap trails represents swimming in four different stroke styles with three levels of speed: starting with slow-speed butterfly style lap, continuing with medium speed butterfly style lap, and at the end finishing with full-speed front crawl style lap. A single 6DoF IMU device (3D accelerometer and 3D gyroscope) captures enough information for accurate swimming style recognition. Backstroke style is easily identified by simply observing A_z accelerometer component, which gives information about the average body position relative to the gravitational vector g_0. The front crawl style can be identified by observing the relevance of rotation around axis $1y$, which is the same as for backstroke. The remaining two styles, butterfly and breaststroke, can be distinguished from the shapes of the signal rotation angle around axis $1x$.

Typical shapes of the body rotation angle signals, derived from the swimming style specific dominant rotation axes, are shown in Fig. 7.35. The swimming coach manually measured swimming phases with a hand stopwatch. The coach counted the number of strokes, noted down the start phase, gliding phase, and the pure swimming

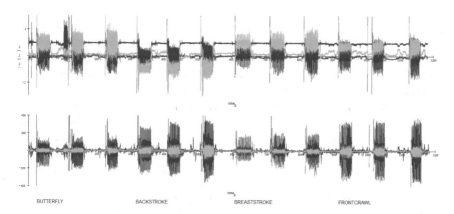

Fig. 7.34 3D accelerometer and 3D gyroscope signals during a 20 min record of the swimming action. A sequence of twelve laps with three levels of speed in four different swimming disciplines: butterfly, backstroke, breaststroke and front crawl. Signals are colour coded: X = red, Y = blue, Z = green (Umek and Kos 2018a, 2018b)

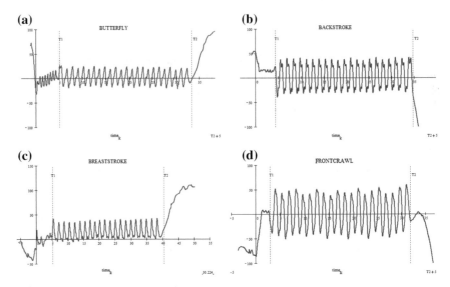

Fig. 7.35 Rotation angle around the dominant rotation axis for four different swimming disciplines. Rotation angle around mediolateral axis (pitch) is coloured red and rotation angle around longitudinal axis (roll) is coloured blue. Results are shown for medium speed trials (Umek and Kos 2018a, 2018b)

phase. The starting times of each phase are defined by time instants: t_0, $t_0 + T_1$, and $t_0 + T_2$. Measured values match the calculated rotation angle signal as shown in Fig. 7.35.

Timing diagrams in Fig. 7.35 offer a valuable information with accurate and detailed parameters in regard to the swimming rhythm (periods and peaks), which cannot be observed by a simple manual measurement techniques. Basic data analysis offers valuable information about the swimming rhythm. Stroke periods and rotation peak values serve as feedback information to the coach. Figure 7.36 shows the first feature extraction steps for a rotation angle signal in the butterfly swimming style. Local maxima, local minima, and zero-crossing signal points are needed to calculate the basic cyclic signal parameters. Figure 7.37 shows stroke period and peak-to-peak pitch angle features extracted from the signal in Fig. 7.35a.

Analysis, similar to the one shown for a butterfly style signal (Figs. 7.36 and 7.37), has been done for all four swimming disciplines. The difference is only in the observed principal rotation axis: the swimming rhythms in butterfly and breaststroke styles are obtained from the rotation angle around axis $1x$, similarly the swimming rhythm in backstroke and front crawl styles are obtained from the rotation angle around axis $1y$. The results of the swimming rhythm parameters calculation for all twelve single-lap trials are collected in Table 7.5.

Parameters from the first six columns in Table 7.5 are acquired manually by the coach during the trial. However, most of them can be detected automatically from the acquired signals:

Table 7.5 Basic swimming rhythm parameters based on dominant axis rotation signals are listed in last four columns (Umek and Kos 2018a, 2018b)

Lap	Style	Speed	FS time T1 [s]	Lap time T2 [s]	Stroke Count	Stroke Period (Mean) [s]	Stroke Period (StDev) [s]	Rotation Angle P2P (Mean) [deg]	Rotation Angle P2P (StDev) [deg]
1	Butterfly	Slow	4.81	35.78	24	1.30	0.03	40.22	1.94
2		Medium	5.10	33.38	24	1.16	0.02	38.75	1.43
3		Fast	4.36	32.29	25	1.10	0.03	38.28	2.64
4	Backstroke	Slow	4.82	38.56	38	1.72	0.04	74.88	3.29
5		Medium	4.39	34.82	42	1.42	0.04	69.49	3.08
6		Fast	4.61	33.23	43	1.26	0.06	64.31	3.01
7	Breaststroke	Slow	5.40	43.92	18	2.11	0.04	36.21	1.44
8		Medium	5.34	40.52	21	1.64	0.06	38.57	2.45
9		Fast	4.66	38.76	24	1.40	0.05	30.38	1.60
10	Front crawl	Slow	3.06	34.75	38	1.66	0.06	101.41	4.78
11		Medium	2.99	31.91	40	1.42	0.03	90.53	4.95
12		Fast	2.58	29.98	45	1.20	0.03	77.63	6.28

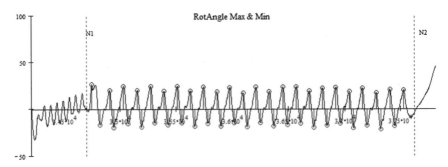

Fig. 7.36 Local maxima and minima detection on cyclic rotation angle signal from Fig. 7.35a

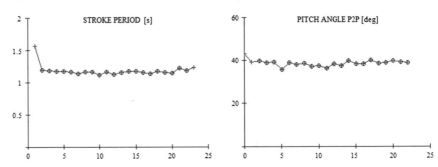

Fig. 7.37 Extracted features based on body rotation signal in butterfly swimming discipline: swimming stroke period and peak-to-peak rotation angle amplitude

- *Swimming style* can be determined directly from the acceleration and gyroscope signals, in fact only three of the six IMU signals are needed: acceleration signal in the anterior-posterior axis and rotation speed signals in the swimming direction axis and medio-lateral direction axis.
- The *starting time* instant $t=0$ can be detected from the relevant swimming style signals shown in Fig. 7.35.
- The time instant T_1, which is marking the moment of the *first breath* after the gliding phase and start of the free swimming phase, can be detected from the relevant swimming style signals shown in Fig. 7.35.
- The total *number of strokes* can be counted by stroke detection from signals in Fig. 7.35.
- The *lap time* cannot be detected from the acquired signals without an additional sensor for wall strike detection.

Kayaking

IMU sensors attached to the paddle give information about the paddle movement, which can be done with or without power transfer from the paddle to the kayak motion. For measuring the paddling rhythm and efficiency, acceleration signal from sensor attached to the kayak is more adequate. The differences between both signals

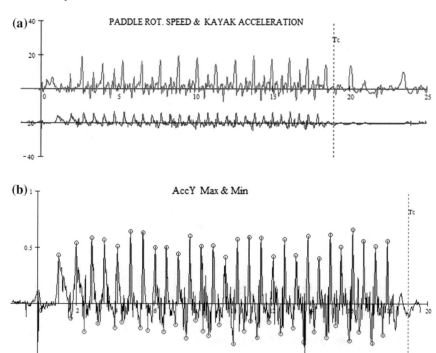

Fig. 7.38 Basic signal selection for stroke rhythm parameters extraction in kayaking

are illustrated by the example shown in Fig. 7.38a. By comparing both signals, it is obvious that the paddle rotation signal is not always correlated with the kayak acceleration signal, that is true after crossing the finish line at $t = T_C$, where the athlete is using the paddle in free run.

The kayak acceleration signal is shown in Fig. 7.38b. The signal includes the vibrations created by the paddler movements and the noise induced by the sliding motion through water. Information about the rhythm is gained by searching the acceleration signal peak values. Analysis of the signal from 0 is similar to the analysis of the swimming strokes in previous subsection. The results for stroke period and acceleration peaks are shown in Fig. 7.39. Identical analysis was conducted on the signals of three different paddlers performing the action at three speed levels: moderate, medium, and fast. Some of the features observed from Fig. 7.39 are: number of strokes $N_{\text{stroke}} = 28$, stroke period $T_{\text{stroke}} = 0.62 \pm 0.08$ s, acceleration maxima $A_{\max} = 0.53 \pm 0.07$ g_0, and acceleration minima $A_{\min} = 0.24 \pm 0.06$ g_0.

Canoeing

In canoeing, similarly as in kayaking, the most relevant sensor is accelerometer attached to the canoe. Figure 7.40 depicts the signal of the medium speed trial. In comparison to the kayaking, canoeing shows stronger pulsation (push-pull) with

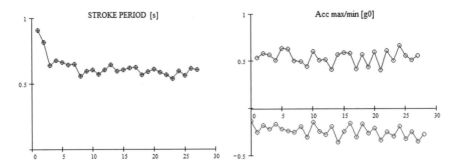

Fig. 7.39 Features extracted from the kayak acceleration signal: stroke period and acceleration peaks

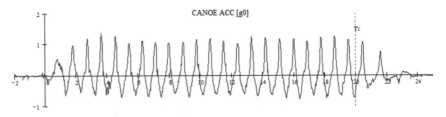

Fig. 7.40 Basis signal for stroke rhythm parameters extraction in canoeing

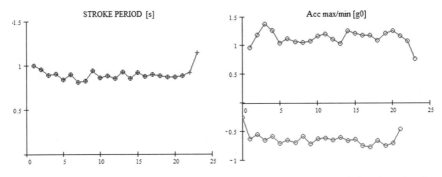

Fig. 7.41 Features extracted from the canoe acceleration signal: stroke period and acceleration peaks

approximately twice larger acceleration and deceleration amplitude. The reason for this effect is larger force with the use of one paddle.

Figure 7.41 summarizes the analysis of stroke period and acceleration peaks of both phases of the stroke. Some of the features, together with their standard deviation, observed from Fig. 7.41 are: number of strokes $N_{stroke} = 22$, stroke period $T_{stroke} = 0.89 \pm 0.04$ s, acceleration maxima $A_{max} = 1.11 \pm 0.17\ g_0$, and acceleration minima $A_{min} = 0.63 \pm 0.15\ g_0$.

7.6.5   Future Development

First tests of the swimming, kayaking, and canoeing application have shown that by using a 6DoF sensor(s), including 3D accelerometer and 3D gyroscope, one can acquire enough information for motion analysis and extraction of the most relevant features of each observed sport.

In swimming we can identify features such as the number of strokes, stroke periods and their variation, underwater swimming phase duration, and others. For a complete automatic analysis with timing information we will add the sensors for the wall strike detection. With the implementation of the above ideas it will be easy to design the final application that will include all the relevant information for the coach. That will allow the coach to concentrate more on other qualitative aspects of training and less on the routine quantitative measurements. Similar observations are true for kayaking and canoeing.

While the results of presented tests were gained through the manual post processing analysis of recorded signals, the coach application will operate with automatic processing and in real time. In practice that means that the coach will be able to see the signals and extracted features in real time - during the athlete's action. Another question is connected to the timing of the presentation of processing results; what is the best time for the coach? This and similar questions will have to be discussed with sport experts or learned through further application tests.

After some time, when the water sport application proves useful for the coaches, the plan it to upgrade it with the user functionality. Such application will give cyclic feedback directly to the athlete. For example, the difference in rotation angles between the left and right arm single swimming stroke or the ratio of power transfer between the paddle and the canoe or kayak. The challenge is not only the real-time data transmission and processing, but also the requirement for waterproof or water resistant devices; from sensors to actuators.

7.7   Swimming Rehabilitation Application

The *Swimming rehabilitation* application is an example of a biofeedback system used in rehabilitation therapy that is based on swimming exercise. The application is aimed at helping patients to properly learn rehabilitation exercises and accelerate their recovery.

7.7.1   Objectives and Functionality

Our primary objective is to develop a group of wearable device based applications for the assistance of prevention and rehabilitation in various fields of healthcare that are related to physical activity. Such applications can be successfully used in healthcare institutions that carry out physical rehabilitation programs (Gruwsved et al. 1996) or with some modifications even by individuals and organizations that practice recreational exercise or sport training (Silva 2014) where their role would be primarily in injury prevention.

Application design includes user and instructor functionality that provide either concurrent or cyclic feedback. The application is used in personal or confined space and it can be implemented in compact or distributed structure.

We have first implemented an application for swimming rehabilitation with a waterproof wearable sensor device including inertial measurement unit and real-time therapist feedback. The swimming rehabilitation application investigates the possible benefits of using IMU-based devices for the support of physical rehabilitation through swimming exercise.

7.7.2   Background

Swimming was chosen to be the first developed biofeedback application for rehabilitation because, historically, hydrotherapy was viewed as a central treatment methodology in the field of physical medicine (Becker 2009), and because numerous research papers show that aquatic therapy and swimming are the exercises of choice in numerous rehabilitation scenarios after injury (Nagle et al. 2017; Prins and Cutner 1999), surgery (Singh et al. 2015) and are beneficial even for people with autism (Yilmaz et al. 2004). In rehabilitation aquatic therapy and swimming, the water buoyancy reduces the effects of gravity on the body. This is particularly beneficial for the spine and joints such as the hips, knees and ankles (Prins and Cutner 1999). Swimming is recognized as an important physical exercise for rehabilitation, because various spinal diseases and injuries show through movement asymmetry during swimming (Becker 2009).

Review papers from Camomilla et al. (2016) and Neiva et al. (2017) present the results of numerous studies about the use of wearable sensors and technology in sport and physical exercise. In Camomilla et al. (2016), the results show that in cyclic sports wearable sensors are most frequently used in distance running and swimming. Neiva et al. (2017) have identified 603 studies published between 2007 and 2016 that discuss wearable technology for measuring physiological and biomechanical parameters. Out of 112 studies that focus on physical activity, 13 were about swimming. Authors of both papers conclude that wearable technology can be effectively used for monitoring physical activities, including swimming.

Guignard et al. (2017) define low-order and high-order behavioural parameters for the analysis of motor control in swimming. While low-order parameters refer to superficial aspects such as position, velocity, and acceleration, high-order parameters refer to dynamics and movement coordination. It has been shown that IMU-based devices can record both low-order and high-order swimming parameters. An important observation of this paper that confirms our system design is that IMU devices have a major advantage over camera-based measurement systems because they can quite easily record continuous data over long time periods and are usually not restricted in space.

Multiple inertial sensors have been used for swimming symmetry assessment in Becker (2009). The system is designed on a microcontroller and a 9DoF inertial sensor. It can record up to 240 s of swimming activity. The swimming data is processed on a personal computer after the exercise. A similar system that was designed as a wrist band including a 3D accelerometer is presented in Delgado-Gonzalo et al. (2016). It can record up to 12 h of swimming activity for post processing and is able to present some parameters on the device LCD display in real time. The shortcoming of these two systems is that the data is available only after the exercise, which makes the improvement process of motor control lengthier.

The wearable biofeedback system presented by Li et al. (2016) has proven effective and reliable for self-training and performance monitoring of recreational swimmers. It uses a real-time biofeedback unit with a sensor, a processing unit and actuators attached to the body of the swimmer. It corresponds to our definition of a user system (see Sect. 4.4). Unfortunately, such a system is not directly usable for rehabilitation purposes because in the early rehabilitation phases, the exercise must be supervised by a therapist.

The inertial measurement system for swimming rehabilitation, presented by Parvis et al. (2017), is capable of sending the swimming signals and data in real time over a WiFi connection. It can measure several swimming parameters. The authors claim that the system is capable of a quasi-on-line self-assessment. The shortcoming of this system is that the swimming parameters are not available to the therapist in real time, that is, during the swimming exercise.

As mentioned in the beginning of this section, swimming has been proven to be a very effective therapy in physical medicine rehabilitation. Our application focuses on the acquisition of as many key performance indicators as possible by using a simple and robust measurement system that can be adopted by therapists without any need for specialized technical support. Most therapists are still using only direct visual information. We are deeply convinced that our application can give them very valuable supplemental information.

Augmented real-time feedback can improve the efficiency of the therapists' work in several ways: (a) therapists are not required to constantly monitor the swimming exercise execution and can focus on other elements of the therapy; (b) therapists can likely work with more people at the same time than without such a system; and (c) technical equipment improves measurement precision, and therapists are given information that they cannot perceive by their senses; for example, swimming exercise consistency, variations in stroke period, movement symmetry, tiredness, and others.

7.7.3 System Architecture and Setup

Because of the specific environment, the monitoring of swimming activity in rehabilitation is traditionally done by the eye of the therapist or by complex and expensive camera-based systems. The eye of the therapist might not notice certain details of the underwater movements, and the camera systems usually give the information about the parameters and the performance of the user only after the exercise has been completed.

Systems based on wearable sensor devices have the potential to overcome both of these shortcomings. It is of crucial importance that wearable sensor devices can be attached to various parts of the body and that they can acquire and (wirelessly) send signals and data of interest to the processing device in real time. The processing, analysis and feedback of the swimming exercise results can be performed in various ways and in different system architectures. We present two different wearable sensor device based applications for monitoring and feedback in swimming rehabilitation.

Multi-user Therapist Application

A multi-user therapist application performs simultaneous monitoring and counselling of several users during their rehabilitation. In a multi-user therapist application, presented in Fig. 7.42, one or more wearable sensor devices are attached to various parts of the users' bodies. Signals and data of interest are wirelessly transmitted to the processing device. While several different wireless technologies can be used for the transmission of signals and data between sensor devices, the processing device, and the therapist, the most suitable are wireless local area network technologies (Kos et al. 2018). The processing device separately processes signals and data from each active user and sends the results to the therapist.

The application offers user-by-user parameters, performance indicators, and time-dependent graphs in real time. When necessary, the therapist gives only the relevant feedback to the supervised users about their performance and possible corrections or improvements of their movement, action, and exercise. The feedback can be given orally or through any other available communication path. For example, users can also have earphones for receiving audio feedback, or they could have attached tactile actuators.

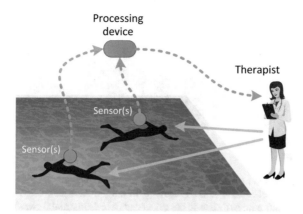

Fig. 7.42 Multi-user therapist application. User exercise is monitored by wearable sensor devices, which wirelessly send signals and data of interest to the nearby processing device. The results of the exercise parameters and user performance are calculated in real-time and are instantly available to the therapist, who gives educated feedback to the users (Kos and Umek 2018)

The advantages of the presented multi-user therapist application are its real-time operation to the therapist with the possibility of immediate therapist feedback to the user. Such feedback can significantly improve and most likely also speed-up the rehabilitation therapy because the therapist has information about incorrect and possibly harmful movements throughout the duration of the exercise. This can be beneficial not only to the user but also to the healthcare system in general. For example, therapy supported by real-time feedback systems can make rehabilitation therapy more effective and shorter, and therefore cheaper.

User Application

User application resembles the compact architecture of the biofeedback system, because all of its elements are attached to the user as shown in Fig. 7.43. In general, sensors, actuators, and processing devices of user application are separate devices. They can also be combined into one device containing all of the above, similarly as in (Umek et al. 2015).

User application operates in real time, giving concurrent feedback to the user. It requires an *educated user*, that is, a user that has been trained in the correct and appropriate use of the application and in the correct and appropriate exercise execution. Both are easily achievable by using the multi-user therapist application in the first phase, when the user will learn the correct execution of the exercise and possible feedback information with the help of a therapist. In the second phase, the user will use the application autonomously, with possible consultation with the therapist.

Therapy with the user application, if implemented appropriately, can offer even greater benefits to the rehabilitation therapy and the healthcare system than the use of the multi-user therapist application alone. The user application allows users to

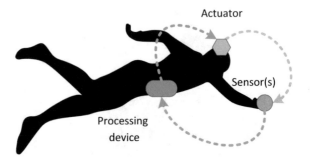

Fig. 7.43 User application. All system devices are attached to the user. Wearable sensor device(s) send signals and data of interest to the processing device on the user's body. The results about the exercise parameters and performance are calculated in real-time and forwarded to the actuator attached to the user's body (Kos and Umek 2018)

Fig. 7.44 The movement of the body is detected by sensors in the belt strapped to the lower back of the user. The belt and the sensor do not interfere with the user during the swimming phase

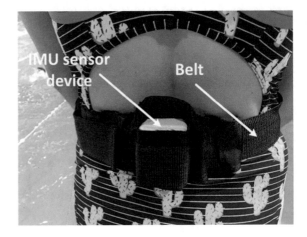

practice previously learnt exercises without the constant supervision of the therapist. In this way, healthcare systems can become more efficient, because users are simultaneously saving therapists' time and advancing in their rehabilitation therapy. Supervision of the appropriate, correct, and safe way of performing the therapeutic exercise is done by the user application operating in real-time.

Application Test Setup

Application tests were done with a waterproof 6DoF sensor, including a 3D accelerometer and 3D gyroscope, inside one wearable sensor device. A similar sensor type and setup can be found in Stamm et al. (2013).

The lower back was selected as the most appropriate placement for the sensor unit, as shown in Fig. 7.44. We have used a custom designed belt that held the sensor device firmly in place, even in high-intensity swimming.

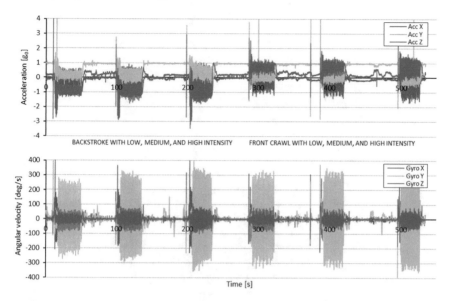

Fig. 7.45 Accelerometer and gyroscope signals during a sequence of six laps in backstroke and front crawl swimming styles and three different levels of swimming intensity: low, medium, and high (Kos and Umek 2018)

7.7.4 Results

Backstroke and front crawl swimming styles are favourable in therapeutic swimming exercise. The swimming test consisted of the continuous recording of three laps in the backstroke swimming style and three laps of the front crawl swimming style with intermediate resting periods. The complete record of accelerometer and gyroscope signals is shown in Fig. 7.45. The sequence of six laps starts with three levels of intensity (low, medium, high) in backstroke style and continues with three levels of intensity in front crawl style.

The recorded test data, performed by an elite professional swimmer, is used to evaluate the swimming motion dynamics of the lower back. As shown in Fig. 7.45, the acceleration magnitude does not exceed $\pm 2\ g_0$, and angular velocity stays in the range of ± 350 dps. Stroke frequency in elite swimming is below 2 Hz, but coordinated leg kicking motion with multiple higher frequencies is detected at this sensor position. The actual acceleration and angular velocity signal spectra are narrower than 50 Hz. Therefore, accelerometer and gyroscope signals are sampled at 100 Hz. These dynamic ranges should not be exceeded when sensors are used by non-professional swimmers during their physical exercises or rehabilitation therapy.

The sensor attached at the lower back near the body's centre of mass is aligned to the body longitudinal axis. For the basic data analysis of the backstroke and front crawl swimming styles, only the gyroscope y-axis signal component is needed. Typical shapes of the body rotation angle around the longitudinal axis are shown

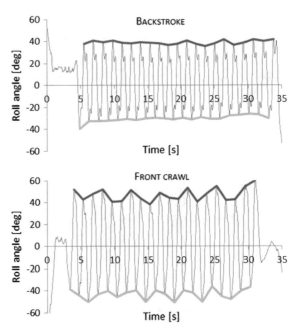

Fig. 7.46 Rotation angle around the longitudinal axis (roll) for backstroke and front crawl swimming styles. The results are shown for medium intensity laps. Maximum and minimum values presented with red and green lines, respectively, give information about basic temporal and spatial swimming parameters (Kos and Umek 2018)

in Fig. 7.46. The detailed timing diagrams acquired from Fig. 7.46 offer valuable information with more accurate and detailed parameters in regard to the swimming rhythm, which cannot be observed by simple manual measurement techniques.

Basic data analysis offers valuable information about the swimming rhythm. For example, stroke periods and rotation peak values serve as feedback information to the coach in elite swimming training, to the therapist in swimming rehabilitation therapy, or directly to the user practicing daily swimming exercise. Figure 7.46 shows the first feature extraction steps for a rotation angle signal in the backstroke and front crawl styles. Local maxima, local minima, and zero-crossing signal points are needed to calculate the basic cyclic signal parameters. Figures 7.47 and 7.48 show peak-to-peak pitch angle and stroke period features extracted from the signals in Fig. 7.46.

Basic spatial and temporal swimming parameters in backstroke and front crawl are extracted from rotations around the longitudinal axis shown in Fig. 7.46. Sequences of "roll" angle signal peaks for all six laps from the second swimming exercise are shown in Fig. 7.47. Low-intensity swimming (green) has a larger angular magnitude than high-intensity swimming (red), which is performed at a higher pace. Collected data point time stamps are used in the stroke period calculation shown in Fig. 7.48. Positive rotation strokes, which represent the path from the blue to the red marker points in Fig. 7.46, are marked with circles in Fig. 7.48, and negative rotation strokes

Fig. 7.47 Peak values in rotation angle around the horizontal plane in swimming direction axis (roll). Extracted features are from six laps of backstroke and front crawl swimming styles, with low, medium and high swimming intensity. Swimming intensity colour codes: green = low, blue = medium, red = high (Kos and Umek 2018)

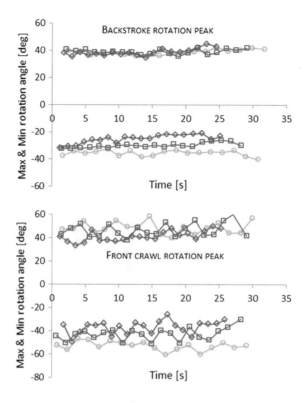

are marked with crosses in Fig. 7.48. Double-stroke periods, which represent the sum of consecutive left-handed and right-handed stroke periods, are marked with diamonds in Fig. 7.48. High-intensity swimming (red) has shorter stroke periods and consequently higher stroke counts than low-intensity swimming (green).

The measured spatial and temporal parameters, angle magnitude and stroke period, change with swimming intensity. The lap-averaged values of the peak–to-peak rotations around the longitudinal axis shown in Fig. 7.47 and the lap-averaged values of the double-stroke periods shown in 7.48 are collected in Table 7.6. Lower intensity swimming is performed with shorter periods and larger rotation magnitudes than higher intensity swimming. The variation in swimming intensity can also be detected during each stroke. For example, the variation in swimming intensity, which can be a result of tiredness, can be detected in real time, before the end of each lap.

Symmetry in swimming motion in backstroke and front crawl can be acquired from the data in Figs. 7.47 and 7.48. The spatial symmetry parameter is related to the rotation angle peak ratio. Asymmetry in the rotation angle magnitude is related to the difference in power used during the left-hand and the right-hand single stroke and to the motor control of the swimmer. The symmetry ratio of the angle peaks in both directions is calculated for backstroke and front crawl. The average lap results are shown in the fourth column of Table 7.6. The temporal symmetry parameter in

Fig. 7.48 Stroke periods are extracted from rotation angle data (roll). Left-handed stroke periods, right-handed stroke periods, and double stroke periods are measured for different swimming styles and swimming intensity levels. Swimming intensity colour codes: green = low, blue = medium, red = high (Kos and Umek 2018)

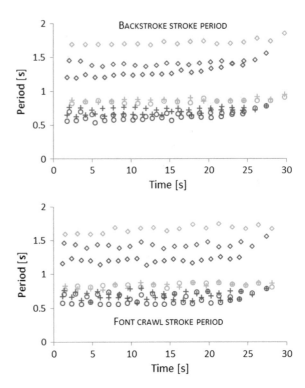

Table 7.6 Basic spatial, temporal, and symmetry ratio parameters (Kos and Umek 2018)

Style - Lap	Peak-to-peak rotation angle [deg]	Double stroke period [s]	Rotation angle symmetry	Stroke period symmetry
Backstroke - 1	74.5	1.72	1.09	0.97
Backstroke - 2	69.1	1.42	1.32	0.90
Backstroke - 3	64.0	1.27	1.59	0.89
Backstroke - 1	101.6	1.67	0.93	1.01
Backstroke - 2	90.5	1.43	1.11	0.96
Backstroke - 3	78.1	1.2	1.64	0.91

backstroke and front crawl can be evaluated from the ratio of left-handed and right-handed stroke periods from the results in Fig. 7.48. The symmetry ratio is calculated for backstroke and front crawl. The average lap results are shown in the fifth column of Table 7.6. Stroke timing asymmetry is much higher during the higher intensity swimming, where the swimmer works with full power and is less precise in their motor control.

7.7.5 Future Development

The presented results of swimming rehabilitation application test have shown that 6DoF inertial sensors provide enough information for motion analysis and extraction of the most relevant features of swimming exercise. It is possible to identify features such as the number of strokes, stroke periods, their variations, symmetry, and others. The swimming rehabilitation application will include all the relevant information for the rehabilitation therapists. This will allow them to concentrate more on other qualitative aspects of rehabilitation and less on the routine quantitative measurements and performance parameters.

For the rehabilitation therapist, it is particularly important that such feedback applications provide the information, which is difficult or impossible to acquire through traditional manual observation techniques. In the case study of swimming rehabilitation, such parameters are, for example, swimming symmetry (Parvis et al. 2016) and detection of tiredness. Both parameters can be accurately measured by our feedback system, and the therapist is able to see the swimming exercise signals and extracted features in real time, that is, during the user's action.

A basic signal processing suffices for simple features, such as signal peaks, stroke symmetry, and stroke rate variation. For some other features, more elaborate signal and data analysis methods are needed, such as pattern recognition and machine learning (Wei et al. 2016). A detailed study of these methods is beyond the scope of this paper.

Our future work in the field of swimming rehabilitation application includes several tasks that will help therapists and users to truly benefit from our real-time feedback systems. The wearable sensor device will be redesigned to have a more hydrodynamic shape and will be easier to mount properly and use with greater comfort. The application will be adapted to be more user and therapist friendly and usable without constant support from technical personnel. A field study including a larger number of users investigating the benefits of augmented real-time feedback will be conducted. Future work will be aimed at determining the greatest effectiveness and usefulness of our feedback concept in swimming rehabilitation therapy.

References

Abdul Razak AH, Zayegh A, Begg RK, Wahab Y (2012) Foot plantar pressure measurement system: A review. Sensors 12(7):9884–9912

Adelsberger R, Aufdenblatten S, Gilgien M, Tröster G (2014) On bending characteristics of skis in use. Procedia Engineering 72:362–367

Ahmadi, A., Destelle, F., Monaghan, D., O'Connor, N. E., Richter, C., & Moran, K. (2014, November). A framework for comprehensive analysis of a swing in sports using low-cost inertial sensors. In SENSORS, 2014 IEEE (pp. 2211–2214). IEEE

Barbosa AC, Castro FDS, Dopsaj M, Cunha SA, Júnior OA (2013) Acute responses of biomechanical parameters to different sizes of hand paddles in front-crawl stroke. J Sports Sci 31(9):1015–1023

Becker BE (2009) Aquatic therapy: scientific foundations and clinical rehabilitation applications. PM&R 1(9):859–872

Betzler NF, Monk SA, Wallace ES, Otto SR (2012) Effects of golf shaft stiffness on strain, clubhead presentation and wrist kinematics. Sports Biomech 11(2):223–238

Camomilla V, Bergamini E, Fantozzi S, Vannozzi G (2016) In-field use of wearable magneto-inertial sensors for sports performance evaluation. In: ISBS-conference proceedings archive, vol 33(1)

Chambers R, Gabbett TJ, Cole MH, Beard A (2015) The use of wearable microsensors to quantify sport-specific movements. Sports Med 45(7):1–17

Choi YC, Kim HK, Shim KB (2016) Analyzing the characteristics of golf driver shafts with using a strain gage. J Ceram Process Res 17:113–117

Chun S, Kang D, Choi HR, Park A, Lee KK, Kim J (2014) A sensor-aided self coaching model for uncocking improvement in golf swing. Multimed Tools Appl 72(1):253–279

Delgado-Gonzalo R, Lemkaddem A, Renevey P, Calvo EM, Lemay M, Cox K, Bertschi M (2016) Real-time monitoring of swimming performance. In: 2016 IEEE 38th annual international conference of the engineering in medicine and biology society (EMBC). IEEE, pp 4743–4746

Dopsaj M, Matković I, Thanopoulos V, Okičić T (2003) Reliability and validity of basic kinematics and mechanical characteristics of pulling force in swimmers measured by the method of tethered swimming with maximum intensity of 60 seconds. Facta Univ Ser: Phys Educ Sport 1(10):11–22

Doyle B (2015) Experts weigh in on head movement during the golf swing. Forever better golf. https://foreverbettergolf.com/articles/experts-weigh-in-on-head-movement-during-the-golf-swing/. Accessed 10 June 2018

Ebling MR (2016) IoT: from sports to fashion and everything in-between. IEEE Pervasive Comput 4:2–4

Falda-Buscaiot T, Hintzy F, Coulmy N (2016) Ground reaction force comparison between both feet during giant slalom turns in alpine skiing. In ISBS-conference proceedings archive, vol 33(1)

Ganzevles S, Vullings R, Beek PJ, Daanen H, Truijens M (2017) Using tri-axial accelerometry in daily elite swim training practice. Sensors 17(5):990

Gruwsved Å, Söderback I, Fernholm C (1996) Evaluation of a vocational training programme in primary health care rehabilitation: a case study. Work 7(1):47–61

Guignard B, Rouard A, Chollet D, Seifert L (2017) Behavioral dynamics in swimming: the appropriate use of inertial measurement units. Front Psychol 8:383

Guo J, Zhou X, Sun Y, Ping G, Zhao G, Li Z (2016) Smartphone-based patients' activity recognition by using a self-learning scheme for medical monitoring. J Med Syst 40(6):140

Hsu YL, Chen YT, Chou PH, Kou YC, Chen YC, Su HY (2016) Golf swing motion detection using an inertial-sensor-based portable instrument. In 2016 IEEE international conference on consumer electronics-Taiwan (ICCE-TW). (pp 1–2)

Jakus G, Stojmenova K, Tomažič S, Sodnik J (2017) A system for efficient motor learning using multimodal augmented feedback. Multimed Tools Appl 76(20):20409–20421

Jensen U, Schmidt M, Hennig M, Dassler FA, Jaitner T, Eskofier BM (2015) An IMU-based mobile system for golf putt analysis. Sports Eng 18(2):123–133

Jiao L, Bie R, Wu H, Wei Y, Kos A, Umek A (2018) Golf swing data classification with deep convolutional neural network. IPSI BGD Trans Internet Res 14(1):29–34

Kirby R (2009) Development of a real-time performance measurement and feedback system for alpine skiers. Sports Technol 2(1–2):43–52

Kos A, Umek A (2018) Wearable sensor devices for prevention and rehabilitation in healthcare: swimming exercise with real-time therapist feedback. IEEE Internet Things J. https://doi.org/10.1109/jiot.2018.2850664

Kos A, Tomažič S, Umek A (2016) Suitability of smartphone inertial sensors for real-time biofeedback applications. Sensors 16(3):301

Kos A, Milutinović V, Umek A (2018) Challenges in wireless communication for connected sensors and wearable devices used in sport biofeedback applications. Future Gener Comput Syst. https://doi.org/10.1016/j.future.2018.03.032

Kunze K, Minamizawa K, Lukosch S, Inami M, Rekimoto J (2017) Superhuman sports: Applying human augmentation to physical exercise. IEEE Pervasive Comput 16(2):14–17

Li R, Cai Z, Lee W, Lai DT (2016) A wearable biofeedback control system based body area network for freestyle swimming. In: 2016 IEEE 38th annual international conference of the engineering in medicine and biology society (EMBC). IEEE, (pp 1866–1869)

Li X, Wang C, Wang H, Guo J (2017) Real-time dynamic data analysis model based on wearable smartband. In: International conference on intelligent and interactive systems and applications. Springer, Cham, (pp 442–449)

Lightman K (2016) Silicon gets sporty. IEEE Spectr 53(3):48–53

Llosa J, Vilajosana I, Vilajosana X, Navarro N, Surinach E, Marques JM (2009) REMOTE, a wireless sensor network based system to monitor rowing performance. Sensors 9(9):7069–7082

Magalhaes FAD, Vannozzi G, Gatta G, Fantozzi S (2015) Wearable inertial sensors in swimming motion analysis: a systematic review. J Sports Sci 33(7):732–745

Mendes JJA Jr, Vieira MEM, Pires MB, Stevan SL Jr (2016) Sensor fusion and smart sensor in sports and biomedical applications. Sensors 16(10):1569

Michahelles F, Schiele B (2005) Sensing and monitoring professional skiers. IEEE Pervasive Comput 4(3):40–45

Mitsui T, Tang S, Obana S (2015 Support system for improving golf swing by using wearable sensors. In: 2015 eighth international conference on mobile computing and ubiquitous networking (ICMU). IEEE, (pp 100–101)

Mooney R, Corley G, Godfrey A, Quinlan LR, ÓLaighin G (2015) Inertial sensor technology for elite swimming performance analysis: a systematic review. Sensors 16(1):18

Nagle EF, Sanders ME, Franklin BA (2017) Aquatic high intensity interval training for cardiometabolic health: benefits and training design. Am J Lifestyle Med 11(1):64–76

Najafi B, Lee-Eng J, Wrobel JS, Goebel R (2015) Estimation of centre of mass trajectory using wearable sensors during golf swing. J Sports Sci Med 14(2):354

Nakazato K, Scheiber P, Müller E (2011) A comparison of ground reaction forces determined by portable force-plate and pressure-insole systems in alpine skiing. J Sports Sci Med 10(4):754

Nam CNK, Kang HJ, Suh YS (2014) Golf swing motion tracking using inertial sensors and a stereo camera. IEEE Trans Instrum Meas 63(4):943–952. [192]

Naruo T, Kawashima K, Kimura T, Oota Y, Kanayama T (2013) Golf swing analysis by an inertia sensor and selecting optimum golf club. In: ISBS-conference proceedings archive, vol 1(1)

Neiva HP, Marques MC, Travassos BF, Marinho DA (2017) Wearable technology and aquatic activities: a review. Motricidade 13(1):219

Nemec B, Petrič T, Babič J, Supej M (2014) Estimation of alpine skier posture using machine learning techniques. Sensors 14(10):18898–18914

Parvis M, Grassini S, Angelini E, Scattareggia P (2016) Swimming symmetry assessment via multiple inertial measurements. In: 2016 IEEE international symposium on medical measurements and applications (MeMeA). IEEE, (pp. 1–6)

Parvis M, Corbellini S, Lombardo L, Iannnucci L, Grassini S, Angelini E (2017) Inertial measurement system for swimming rehabilitation. In: 2017 IEEE international symposium on medical measurements and applications (MeMeA). IEEE, (pp. 361–366)

Prins J, Cutner D (1999) Aquatic therapy in the rehabilitation of athletic injuries. Clin Sports Med 18(2):447–461

Qualisys, Motion Capture System. http://www.qualisys.com. Accessed 27 June 2018

Sakurai Y, Fujita Z, Ishige Y (2016) Automatic identification of subtechniques in skating-style roller skiing using inertial sensors. Sensors 16(4):473

Shyr TW, Shie JW, Jiang CH, Li JJ (2014) A textile-based wearable sensing device designed for monitoring the flexion angle of elbow and knee movements. Sensors 14(3):4050–4059

Silva ASM (2014) Wearable sensors systems for human motion analysis: sports and rehabilitation. Doctoral dissertation, Universidade do Porto, Portugal

Singh R, Stringer H, Drew T, Evans C, Jones RS (2015) Swimming breaststroke after total hip replacement; are we sending the correct message. J Arthritis 4(147):2

Stamm A, James DA, Thiel DV (2013) Velocity profiling using inertial sensors for freestyle swimming. Sports Eng 16(1):1–11

Stančin S, Tomažič S (2013) Early improper motion detection in golf swings using wearable motion sensors: The first approach. Sensors 13(6):7505–7521

Sturm D, Yousaf K, Eriksson M (2010) A wireless, unobtrusive kayak sensor network enabling feedback solutions. In 2010 international conference on body sensor networks (bsn). IEEE, (pp 159–163)

Sun Y, Song H, Jara AJ, Bie R (2016) Internet of things and big data analytics for smart and connected communities. IEEE Access 4:766–773

Tessendorf B, Gravenhorst F, Arnrich B, Tröster G (2011) An imu-based sensor network to continuously monitor rowing technique on the water. In 2011 seventh international conference on intelligent sensors, sensor networks and information processing (ISSNIP). IEEE, (pp 253–258)

Ueda M, Negoro H, Kurihara Y, Watanabe K (2013) Measurement of angular motion in golf swing by a local sensor at the grip end of a golf club. IEEE Trans Human-Mach Syst 43(4):398–404

Umek, A., & Kos, A. (2018a). Smart equipment design challenges for real time feedback support in sport. Facta Universitatis, Series: Mechanical Engineering

Umek A, Kos A (2018b) Wearable sensors and smart equipment for feedback in watersports. Procedia Comput Sci 129:496–502

Umek A, Tomažič S, Kos A (2015) Wearable training system with real-time biofeedback and gesture user interface. Pers Ubiquit Comput 19(7):989–998

Umek A, Zhang Y, Tomažič S, Kos A (2017) Suitability of strain gage sensors for integration into smart sport equipment: A golf club example. Sensors 17(4):916

Wang Z, Wang J, Zhao H, Yang N, Fortino G (2016) CanoeSense: monitoring canoe sprint motion using wearable sensors. In: 2016 IEEE international conference on systems, man, and cybernetics (SMC) IEEE

Wei Y, Jiao L, Wang S, Bie R, Chen Y, Liu D (2016) Sports motion recognition using MCMR features based on interclass symbolic distance. Int J Distrib Sens Netw 12(5):7483536

Woods T (2009) Maintain a quiet head http://www.golfdigest.com/golf-instruction/2009-10/tiger_woods_keep_quiet_head. Golf digest. Accessed 26 June 2018

Yilmaz I, Yanardag M, Birkan B, Bumin G (2004) Effects of swimming training on physical fitness and water orientation in autism. Pediatr Int 46(5):624–626

Yu G, Jang YJ, Kim J, Kim JH, Kim HY, Kim K, Panday SB (2016) Potential of IMU sensors in performance analysis of professional alpine skiers. Sensors 16(4):463

Zhang Z, Zhang Y, Kos A, Umek A (2017) A sensor-based golfer-swing signature recognition method using linear support vector machine. Elektrotehniski Vestnik 84(5):247–252

Index

A

Accelerometer, 8, 10, 52, 62–67, 77, 83–87, 89–99, 109, 114, 121, 125, 129–132, 136–138, 140, 141, 143, 144, 147, 148, 152, 158, 161, 165, 167, 169, 172, 173

Accelerometer bias, 84, 85, 87, 95, 99

Accelerometer noise, 94, 96

Accuracy, 6, 8, 11, 17, 28, 42, 51, 61, 62, 64, 65, 81, 86, 89, 90, 94, 95, 100, 101, 130, 141–143

Allan variance, 87–91

Analysis time window, 8, 14

Angle Random Walk (ARW), 88–91, 93, 96, 98

Audio feedback signal, 53, 123

Auditory feedback, 9, 10, 31, 119, 151

B

Bend sensors, 145–147, 152–154

Bias compensation, 81, 85, 90–96, 105, 125

Bias error, 84, 87–89, 92–95

Bias variation, 91, 97, 99

Biofeedback, 1–10, 12–15, 17, 19, 20, 25–32, 35–37, 39–46, 49–53, 55, 56, 58, 61, 62, 64–69, 71–74, 76–78, 81–84, 94, 106–108, 111, 112, 117–122, 124–127, 133, 138, 139, 143, 144, 151, 156, 168, 169

Biofeedback applications, 6, 8, 12, 14, 15, 20, 27, 51, 61, 64, 65, 72, 77, 81–83, 92, 94, 95, 99, 105, 111, 112, 117–120, 126, 156

Biofeedback loop delay, 50, 57, 58, 73

Biofeedback processing device, 6, 9, 41–43, 50, 52, 58, 70

Biofeedback sensors, 29

Biofeedback system, 1, 3, 7–9, 16, 19, 20, 35, 36, 41–46, 49, 50, 52–58, 61, 64, 68–71, 73–77, 82, 105, 106, 109–112, 117, 118, 120, 127, 143, 145, 167, 169, 171

Biofeedback system architecture, 20, 52, 56–58, 68, 106, 111, 117, 118

Biofeedback system physical extent, 52, 53, 55, 57, 117

Biofeedback system timing, 45

Biomechanical biofeedback, 1, 3, 5, 6, 9–14, 16, 20, 26–30, 32, 34, 39, 44, 46, 61, 120

Biomechanical biofeedback system, 3, 8, 9, 11, 41, 121

Biomechanical feedback system classifications, 20

Bluetooth, 42, 72, 108, 111, 112, 114, 118, 130, 132, 135

Body Area Network (BAN), 73, 106, 111

Body Sensor Network (BSN), 71, 72

C

Carving turn, 144, 145, 149, 150, 154

Cloud system architecture, 53

Communication, 12, 16, 17, 41–44, 49, 50, 52–55, 57, 58, 61, 66, 68, 69, 71–78, 81, 105–108, 111, 118, 119, 130, 135, 157, 170

Communication delay, 6, 51, 58, 75

Communication in RT system, 16

Communication protocols, 108, 109

Communication technologies, 16, 17, 20, 44, 51, 72, 81, 105, 107, 110, 111

Concurrent and cyclic biofeedback, 72

© Springer Nature Switzerland AG 2018
A. Kos and A. Umek, *Biomechanical Biofeedback Systems and Applications*,
Human–Computer Interaction Series, https://doi.org/10.1007/978-3-319-91349-0

Printed in the United States
By Bookmasters